# GIS Tutorial for

# ArcGIS®

# Desktop

# 10.8

Wilpen L. Gorr | Kristen S. Kurland

# GIS Tutorial for
# ArcGIS® Desktop
# 10.8

Esri Press
REDLANDS | CALIFORNIA

# Contents

# Part II  Working with spatial data

## Chapter 4  File geodatabases   133

## Chapter 5  Spatial data   159

## Chapter 6  Geoprocessing   185

## Chapter 7  Digitizing   227

# Preface

*GIS Tutorial for ArcGIS® Desktop 10.8* is an update to GIS Tutorial 1: Basic Workbook 10.3.x, which was the direct result of the authors' experience in teaching GIS to undergraduate and graduate students in several departments and disciplines at Carnegie Mellon University in Pittsburgh, Pennsylvania, and to high school students in a summer program at Carnegie Mellon University, as well as to professionals in workshops.

*GIS Tutorial*, seventh edition, is a hands-on workbook with step-by-step exercises that take the reader from the basics of using ArcMap™, part of ArcGIS Desktop, through performing many kinds of spatial analyses. Instructors can use this book for the lab portion of a GIS course, or individuals can use it for self-study. You can learn a lot about GIS concepts and principles by doing, and we provide many short notes on a just-in-time basis to help this kind of learning.

The book has three parts. Part I, "Using and making maps," is essential for all beginning students. The chapters of part II, "Working with spatial data," instruct students to download spatial data from government sources on the internet and prepare the data for use in GIS. Part III, "Analyzing spatial data," puts GIS to work for more than mapping, introducing many analytical tools that are unique to GIS. The parts of the book are largely independent of each other, and you can use them in the order that best fits your needs or your class's needs. Moreover, while it's best to cover the chapters of part I in order, you can use the chapters of parts II and III in any order.

Part I has three chapters. In chapter 1, readers learn the basics of working with existing GIS data and maps. Chapter 2 instructs the reader to build many of the kinds of maps commonly used in practice. Chapter 3 covers the forms in which maps are used, including static map layouts, interactive maps, and map animations.

With its five chapters, Part II is the longest and provides much expertise on building a GIS. The exercises in chapter 4 teach readers how to create geodatabases and import data into them, join tables, and carry out other data processing steps. Chapter 5 explores the basic spatial data formats used in GIS, shows readers how to download basemaps and related data from major governmental spatial data suppliers, and then how to prepare and import downloaded spatial data into ArcGIS. Chapter 6 covers geoprocessing tools and workflows needed to transform basemaps into study areas in a GIS. Included is an exercise on creating macros for automating multiple-step workflows. Editing spatial data is a vital part of GIS work, so chapter 7 teaches how to digitize and edit vector data and transform data to match geographic coordinates. In chapter 8, students learn how to transform textual address data into mapped points through the geocoding process.

Part III has four chapters on analyzing spatial data. Chapter 9 focuses on standard tools for spatial analysis, including buffers for proximity analysis, spatial joins, and intersections for selecting areas that meet multiple criteria. Chapters 10–12 introduce three ArcGIS extensions. Chapter 10 introduces ArcGIS® 3D Analyst™, enabling students to create 3D scenes, conduct fly-through animations, and conduct landform analysis. Chapter 11 introduces ArcGIS® Spatial Analyst™ for creating and analyzing raster maps, including density maps and site suitability analysis. Finally, chapter 12 introduces ArcGIS® Network Analyst™, including solving the "traveling salesperson" routing problem, building network datasets from TIGER/Line files, creating travel-time polygons for proximity analysis, and solving the multiple-vehicle routing problem such as for deliveries from multiple trucks and multiple warehouses.

To reinforce the skills learned in the step-by-step exercises and provoke critical problem-solving skills, short Your Turn assignments are included throughout each chapter, with challenging homework assignments at the end of most chapters. The quickest way to increase GIS skills is to follow up step-by-step instructions with independent work, and the assignments provide these important learning components. Note that you must complete most Your Turn assignments to work on subsequent exercise steps.

## Downloading the data and software

This book comes with downloadable exercise and assignment data, completed exercises in the EsriPress\GIST1\MyExercises\FinishedExercises folder, and an evaluation code that can be used to download a free trial of ArcGIS Desktop 10.8.1 Advanced software.

You will need the authorization code found on the inside back cover of this book to download the software trial, at go.esri.com/GISTDesktopEVATrial. Available to anyone is a 14-day trial, at **esri.com/arcgis/trial**.

To download the data for the tutorials and exercises, go to ArcGIS Online, at **go.esri.com /GISTDesktopData**. The data is in the Learn ArcGIS organization's GIS Tutorial for Desktop 10.8 (Esri Press) group.

For updates related to this book, go to the book resource page, at **go.esri.com /GISTDesktopResources**.

# Acknowledgments

We would like to thank all who made this book possible.

We have taught GIS courses at Carnegie Mellon University since the late 1980s, always with lab materials we had written. With the feedback and encouragement of students, teaching assistants, and colleagues, we eventually wrote what became this book. We are forever grateful for that encouragement and feedback.

Faculty members of other universities who have taught GIS using previous iterations of this book also have provided valuable feedback. They include Luke Ward of Rocky Mountain College; Irene Rubinstein of Seneca College; An Lewis of the University of Pittsburgh; George Tita of the University of California, Irvine; Walter Witschey of Longwood University; and Jerry Bartz of Brookhaven College.

We are grateful to the many public servants and vendors who generously have supplied us with interesting GIS applications and data, including Kevin Ford of Facilities Management Services, Carnegie Mellon University; Barb Kviz of the Green Practices program, Carnegie Mellon University; Mike Homa of the Department of City Planning, City of Pittsburgh; staff members of the New York City Department of City Planning; Michael Radley of the Pittsburgh Citiparks Department; Pat Clark and Traci Jackson of Jackson Clark Partners; staff of the Pennsylvania Resources Council; former mayor Robert Duffy of the City of Rochester, New York; Kirk Brethauer of the Southwestern Pennsylvania Commission; and employees of several spatial data vendors (including InfoUSA, i-cubed, Information Integration & Imaging LLC, TomTom MultiNet, and Tele Atlas).

Finally, we are much indebted to the wonderful staff at Esri Press for their editorial expertise and beautiful design work and the efficient production and distribution of our book. It's a pleasure working with these dedicated and talented professionals.

# 1

# Introduction

This first chapter familiarizes you with some of the basics of the ArcGIS for Desktop software as well as the general field of GIS. You'll work with map layers and underlying attribute data tables for US states, cities, counties, and streets. These layers are made up of spatial vector features consisting of points, lines, and polygons. Each geographic feature has a corresponding data record, and you'll work with both features and their data records. Vector map layers and raster images are increasingly found in ArcGIS Online, at ArcGIS.com, and on other cloud websites. You will learn more about ArcGIS Online in chapter 3 and about other sources for GIS layers in chapter 5.

Learning objectives

- Open and save a map document
- Work with map layers
- Navigate in map documents
- Measure distances
- Work with feature attributes
- Select features
- Change selection properties
- Work with attribute tables
- Label features

## TUTORIAL 1-1

# Opening and saving a map document

Esri offers three licensing levels of ArcGIS for Desktop, each with increasing capabilities: Basic, Standard, and Advanced. ArcMap, ArcCatalog, ArcScene, and ArcGlobe—all of which you will use in this book—are all part of ArcGIS for Desktop.

## Launch ArcMap

**1** On the Windows taskbar, click Start > ArcGIS > ArcMap 10.8.x. Depending on your operating system and how ArcGIS and ArcMap have been installed, you may have a different navigation menu.

**2** In the resulting ArcMap - Getting Started window, click Existing Maps > Browse for more.

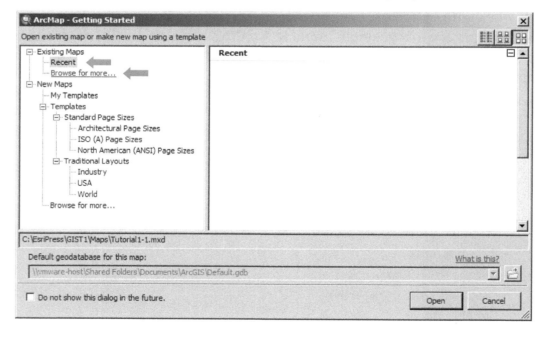

# Open an existing map document

**1** On the Open ArcMap Document dialog box, browse to the drive where you installed the exercise data (for example, C:\EsriPress\GIST1\Maps\).

**2** Click the Tutorial1-1.mxd (or Tutorial1-1) icon.

**3** Click Open. The Tutorial1-1.mxd map document opens in ArcMap, showing a map consisting of the US States layer (zoomed to the 48 contiguous states). The US Cities layer (not yet turned on) is the subset of cities with populations greater than 300,000. The left panel of the ArcMap window is the Table Of Contents window. It serves as a legend for the map as well as several other uses you will learn about. Note that the Tools toolbar, which is floating on the bottom of the screen on the next figure, may be docked somewhere in the interface. If you want, you can anchor it by clicking in its top area, dragging it to a side or the top of the map display window, and releasing when you see a thin rectangle appear. If you do not see the Tools toolbar at all, click Customize > Toolbars > Tools to make it visible. You will learn to use many of the tools in this toolbar throughout this book.

## Save the map document to a new location

You will save all files that you modify or create while working through the tutorials in this book in the \EsriPress\GIST1\MyExercises chapter folders.

**1**  On the Menu bar, click File > Save As.

**2**  Save your map document as Tutorial1-1.mxd to the Chapter1 folder of MyExercises.

## TUTORIAL 1-2
# Working with map layers

Map layers are references to vector data sources such as points, lines, and polygons; raster images; and so forth, representing spatial features that can be displayed on a map. ArcMap displays map layers from a map document such as Tutorial1-1.mxd, but the map document does not contain copies of the map layers. The map layer files remain external to the map document wherever they exist on computer storage media, whether on your computer, a local area network, or on an internet server. Next, you will use the map document's Table Of Contents window to display the map layers in the document.

## Turn a layer on and off

Before GIS existed, mapmakers drew separate layers on clear plastic sheets and then carefully stacked the sheets to make a map composition. Now with GIS, working with layers is much easier.

**1**  **Save your map document as** Tutorial1-2.mxd **to the Chapter1 folder.**

**2**  **In the Table Of Contents window, select the small check box to the left of the US Cities layer.** A check mark appears if the layer is turned on. Note that if the table of contents accidentally closes, you can click Windows > Table Of Contents to reopen it.

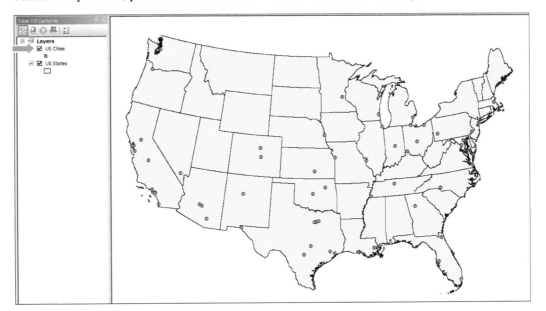

**3** In the Table Of Contents window, clear the check box to the left of the US Cities layer to turn the layer off.

## Add and remove map layers

You can add map layers to the table of contents from their storage locations.

**1** Click the Add Data button ⊕.

**2** On the Add Data dialog box, click the Connect to Folder button 📁.

**3** On the Connect To Folder dialog box, browse to Computer, click the drive where you installed the GIST 1 data (for example, C:\), and click OK. You need to browse to the root drive only and not beyond this. Once this connection is made, you can easily browse to any folder on this computer drive. If your data is stored on an external drive such as a jump drive, you will need to connect to that drive (for example, E:\).

**4** On the Add Data dialog box, browse to \EsriPress\GIST1\Data\UnitedStates.gdb, and click the COCounties layer > Add. ArcMap randomly picks a color for the Colorado counties layer. You will learn how to change the color and other layer symbols later. You may receive a warning about the map's geographic coordinate system.

**5** **Close the Geographic Coordinate Systems Warning window.** ArcMap will place the new layer with Colorado counties correctly over the state of Colorado because all map layers have coordinates tied to the earth's surface. **Note:** Ignore step 5 if you did not get a warning in step 4.

**6** **In the Table Of Contents window, right-click COCounties and click Remove.** This action removes the map layer from the map document but does not delete it from its original storage location, UnitedStates.gdb.

## Use relative paths

When you add a layer to a map, ArcMap stores its path in the map document. When you open a map, ArcMap locates the layer data it needs using these stored paths. If ArcMap cannot find the data for a layer, the layer still will appear in the ArcMap table of contents, but of course it will not appear on the map. Instead, ArcMap places a red exclamation mark (!) next to the layer name to indicate that its path needs repair. You can view information about the data source for a layer and repair it by clicking the Source tab in the layer's property window.

Paths can be absolute or relative. An example of an absolute path is C:\EsriPress\GIST1 \Data\ UnitedStates.gdb\USCities. To share map documents saved with absolute paths, everyone who uses the map document must have the same paths to map layers on their computer. Instead, the relative path option is favored.

Relative paths in a map specify the location of the layers relative to the current location of the map document on disk (.mxd file). Because relative paths do not contain drive letter names, they enable the map and its associated data to point to the same directory structure regardless of the drive or folder in which the map resides. If a map document and associated folders are moved to a new drive, ArcMap will still be able to find the maps and their data by traversing the relative paths.

**1** On the Menu bar, click File > Map Document Properties. Notice the Store relative pathnames to data sources option is checked. This option is for the current map document only.

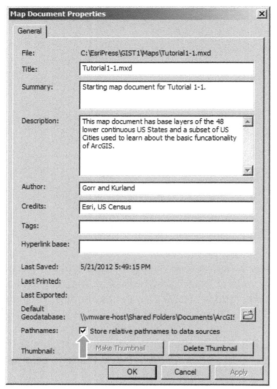

**2** Click OK.

Next, you will verify that all future map documents will use relative paths.

**3** On the Menu bar, click Customize > ArcMap Options > General tab.

**4** On the ArcMap Options dialog box, in the General section, verify that the Make relative paths the default for new map documents option is checked. Click OK.

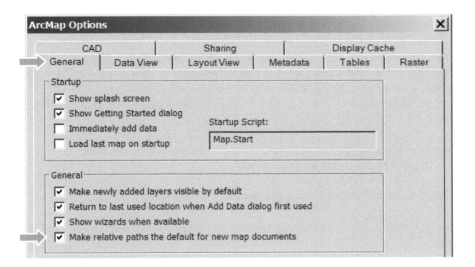

## Drag a layer from the Catalog window

The Catalog window, with its many utility functions, allows you to explore, maintain, and use GIS data. From Catalog, you drag a map layer into the table of contents as an alternative method of adding data.

**1**  On the Menu Bar, click Windows > Catalog.

**2**  In the Catalog window, browse to UnitedStates.gdb in the Data folder.

**3**  Drag COCounties from Catalog to the top of the Table Of Contents window. The map layers in the table of contents draw in order from the bottom up, so if you dropped COCounties below US States, US States covers COCounties. If COCounties is covered, remove it and drag it again from Catalog, this time above US States.

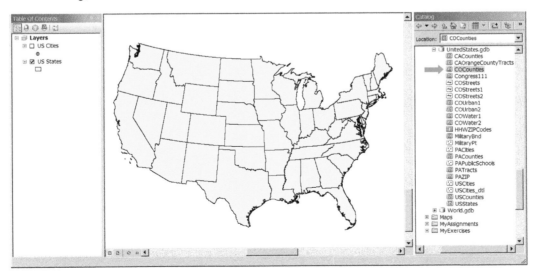

## Use Auto Hide for the Catalog window

Notice that when you opened the Catalog window, it may have opened in pinned-open mode, which keeps the window open and handy for use but covers part of your map. The Auto Hide feature of this application window and of other application windows (such as Table Of Contents and Search) keeps the windows available for immediate use but hides them in between uses, so you have more room for your map.

**1** **Click the Auto Hide button** 🔲 **on top of the Catalog window.** The window closes but leaves a Catalog tab on the right side of the ArcMap window.

**2** **Click the Catalog button** 🔲 Catalog . The Catalog window opens.

Next, you simulate having completed a Catalog task by clicking the map document. The window automatically hides.

**3** **Click any place on the map or table of contents.**

You can pin the window open again, which you do next.

**4** **Click the Catalog button and click the Unpinned Auto Hide button** 🔲 . This pins the Catalog window open until you click the pin again to automatically hide or close the window. Try clicking the map or table of contents to see that the Catalog window remains open.

**5** **Close the Catalog window.**

---

### YOUR TURN

**Use the Add Data or Catalog button to add COStreets, also found in UnitedStates.gdb. These are street centerlines for Jefferson County, Colorado. You may have difficulty seeing the streets, because they occupy only a small area of the map (look carefully above the center of Colorado). Later, in tutorial 1-3, you will learn how to zoom in for a closer look.**

---

## Change a layer's display order

Next, you change the drawing order of layers. For this you must have the List By Drawing Order button selected.

**1** **In the Table Of Contents window, click the List By Drawing Order button** 🔲 **and turn on the US Cities layer.**

**2**   **Drag the US Cities layer to the bottom of the table of contents.** Because ArcMap draws the US Cities layer first now, the US States and Counties layers cover its point markers.

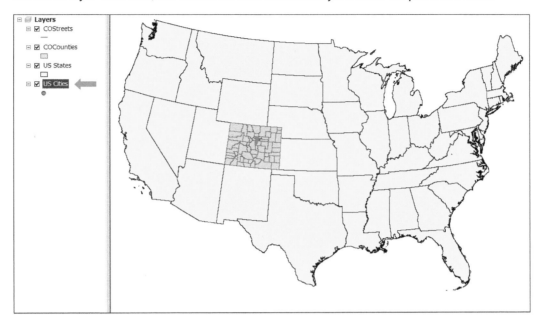

**3**   **Drag the US Cities layer to the top of the table of contents.** ArcMap now draws the US Cities layer last, so you can see its points again.

## Change a layer's color

One of the most helpful capabilities of ArcGIS is how easy it is to change colors and other symbols of layers. First you change the color fill of a layer's polygons.

**1**   **In the Table Of Contents window, click the COCounties layer's legend symbol.** The legend symbol is the rectangle below the layer name in the table of contents.

**2** On the resulting Symbol Selector dialog box, click the Fill Color button ⬜⬛▾ .

**3** In the color palette, click the Tarragon Green tile (column 6, row 5).

**4** Click OK.

## Change a layer's outline color

**1** In the Table Of Contents window, click the layer's legend symbol for COCounties.

**2** On the Symbol Selector dialog box, click the Outline Color button ⬛▾ .

**3** In the color palette, click the Black tile and click OK.

> **YOUR TURN**
> Change the color of the COStreets layer, choosing a light shade of gray (20 percent). You will see the results later. Save your map document.

## TUTORIAL 1-3
# Navigating in a map document

If you are zoomed out as far as possible in a map document, seeing the entire map, this is known as full extent. You can zoom in to any area of the map, resulting in that area filling the map window, giving you a close-up view. The current view of the map is its current extent. You can zoom out, pan, and use several additional means of moving around in your map document. Under the Windows menu, these include the Magnifier window for close-up views without zooming in, the Overview window that shows where you are on the full map when zoomed in, and spatial bookmarks for saving a map extent for future use.

## Zoom in

**1**   **Save your map document as** Tutorial1-3.mxd **to the Chapter1 folder.**

**2**   **On the Tools toolbar, click the Zoom In button** ⊕ **.**

**3**   **Click and hold the mouse button on a point above and to the left of the state of Florida.**

**4**   **Drag the pointer to the bottom and to the right of the state of Florida and release the mouse button.** The process you performed in steps 2 and 3 is called dragging a rectangle.

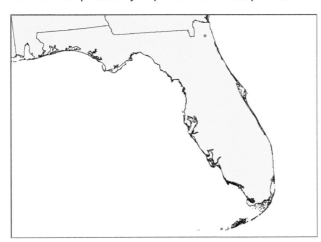

## Fixed zoom in and zoom out

These tools are an alternative for zooming in by fixed amounts.

**1**   **On the Tools toolbar, click the Fixed Zoom In button** ⬚⬚ **.** This zooms in a fixed distance on the center of the current display.

**2**   **Click the map to zoom in, centered on the point you pick.**

**3**   **On the Tools toolbar, click the Fixed Zoom Out button** ⬚⬚ **.** This zooms out a fixed distance from the center of the current zoomed display.

## Pan

Panning shifts the current display in any direction without changing the current scale.

**1**   **On the Tools toolbar, click the Pan button** 🖑.

**1**

2   Move the pointer anywhere into the map view.

3   Clicking the left mouse button, drag the pointer in any direction.

4   Release the mouse button.

## Full, previous, and next extent

The following steps introduce tools that navigate through views you have already created:

1   On the Tools toolbar, click the Full Extent button ⊙. This zooms to a full display of all layers, regardless of whether they are turned on or turned off. Notice that this zooms to all US states and territories.

2   On the Tools toolbar, click the Go Back to Previous Extent button ◀. This returns the map display to its previous extent.

3   Continue to click this button to step back through all of the views.

4   On the Tools toolbar, click the Go to Next Extent button ➡. This moves forward through the sequence of zoomed extents you have viewed.

5   Continue to click this button until you reach the contiguous 48 states.

### YOUR TURN
On the Menu bar, click Customize > ArcMap Options and explore how to control zoom and pan options with your mouse wheel.

## Zoom to layer

Map layers have their own extents and you can zoom to each layer's extent.

1   In the Table Of Contents window, right-click the COCounties layer and click Zoom To Layer. This zooms to the extent of the counties in Colorado.

### YOUR TURN
Zoom to the extent of the layer COStreets (streets in Jefferson County, Colorado). Leave your map zoomed in to the streets.

# Open the Magnifier window

The Magnifier window works like a magnifying glass. As you pass the window over the map display, you see a magnified view of the location under the window.

**1**  On the Menu bar, click Windows > Magnifier.

**2**  Drag the Magnifier over an area of the map to see crosshairs for area selection, and then release to see the zoomed details.

**3** Drag the Magnifier window to a new area to see another detail on the map.

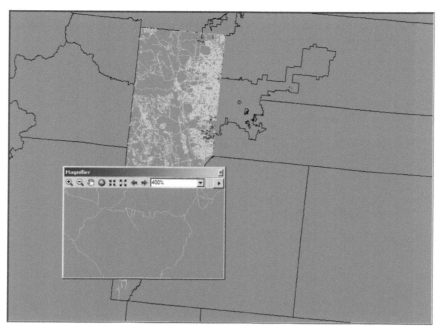

## Change the Magnifier properties

**1** Right-click the title bar of the Magnifier window and click Properties.

**2** Change the Magnify By percentage to 800% if it is not already at that power, and click OK.

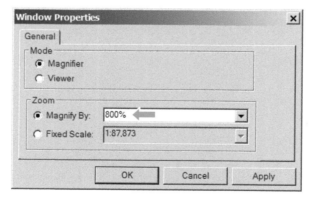

**3** Drag the Magnifier window to a different location and see the resulting view.

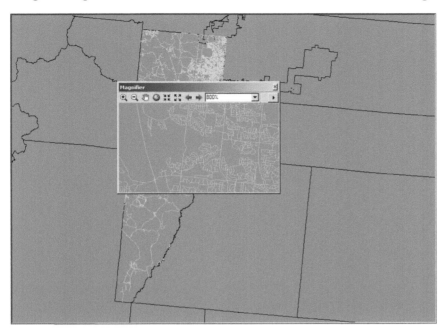

**4** Close the Magnifier window.

## Use the Layers Overview window

The Layers Overview window shows the full extent of the layers in a map. A box in the Layers Overview window shows the currently zoomed area. You can move the box to pan the map display. You can also make the box smaller or larger to zoom the map display in or out.

**1** Zoom to the previous extent until you see the 48 contiguous states.

**2**  Zoom to a small area of the map in the northwest corner of the United States (two or three complete states).

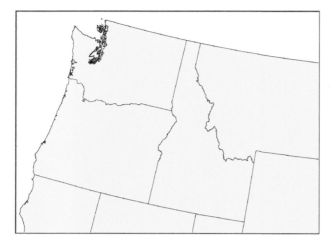

**3**  On the Menu bar, click Windows > Overview. The Overview window shows the current extent of the map highlighted with a red rectangle.

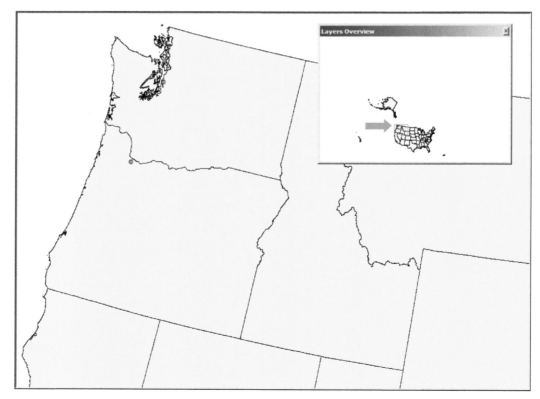

**4** **Move the pointer to the center of the red box and drag to move it to the northeast corner of the United States.** The extent of the map display updates to reflect the changes made in the Layers Overview window. Note that if you right-click the top of the Layers Overview window and click Properties, you can modify the display, including the reference layer for map extents.

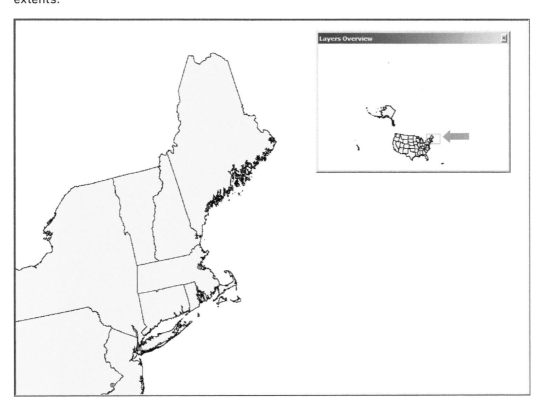

**5** Close the Layers Overview window.

> ## YOUR TURN
> Practice using the Overview window using the Colorado Counties and Colorado Streets layers. In the Overview Properties window, set the reference layer to COCounties. Close the Layers Overview window.

## Use spatial bookmarks

Spatial bookmarks save the extent of a map display so you can return to it later without having to use pan and zoom tools.

**1**   In the Table Of Contents window, turn off the layers COCounties and COStreets.

First, use a bookmark that was already created.

**2**   On the Menu bar, click Bookmarks > 48 Contiguous States.

**3**   Zoom to the state of Florida.

Next, create a new bookmark.

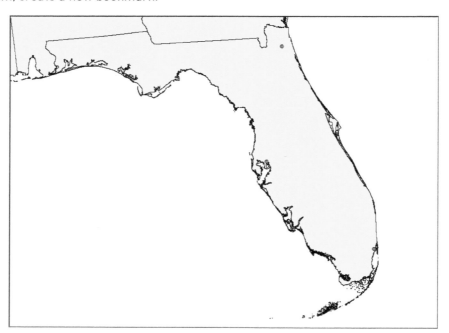

**4**   Click Bookmarks > Create Bookmark, and type Florida in the Bookmark Name field.

**5**   Click OK.

**6**   Click Bookmarks > 48 Contiguous States.

**7**   Click Bookmarks > Florida.

**YOUR TURN**

Create spatial bookmarks for the states of California, New York, and Texas. Try out your bookmarks. Save your map document.

## TUTORIAL 1-4
# Measuring distances

Maps have coordinates enabling you to measure distances along paths that you choose with your mouse and pointer.

## Change measurement units

While a map's coordinates are in specific units such as feet or meters, you can set the measurement tool to gauge distances in any relevant units.

**1**  **Save your map document as** Tutorial1-4.mxd **to the Chapter1 folder.**

**2**  **Click Bookmarks > 48 Contiguous States, and then zoom to the state of Washington.** Washington is the uppermost western state of the 48 contiguous states.

**3**  **On the Tools toolbar, click the Measure button.**

The Measure window appears with the Measure Line tool enabled. The current map units are meters, but miles are more familiar in the United States, so change the units to miles.

**4**  **In the Measure window, click the Units button.**

**5**  **Click Distance > Miles, and leave the Measure window open.**

# Measure the width of Washington State

**1**   Click the westernmost boundary of the state of Washington.

You do not need to match the exact selections made in the following image. Any measurement will demonstrate the procedure.

**2** **Move the mouse in a straight line to the eastern boundary of Washington, and then click the edge.** The distance should be around 335 miles.

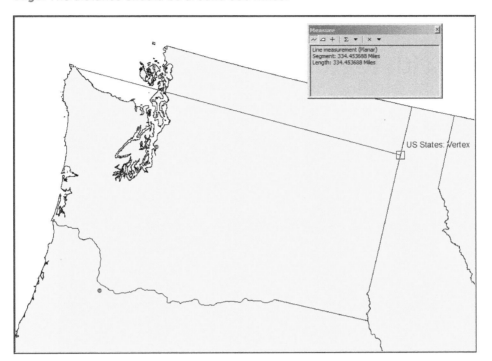

**3** Close the Measure window.

## YOUR TURN

Zoom to the 48 contiguous states and measure the north–south distance of the continental United States from the southern tip of Texas to the northern edge of North Dakota. This distance is approximately 1,600 miles. Measure the east–west distance of the continental United States from Washington to Maine. This distance should be approximately 2,500 miles. Zoom to and measure the north-south distance (top to bottom) of the state of Washington. The distance should be roughly 250 miles. Close the Measure window when finished. Save your map document.

## TUTORIAL 1-5
# Working with feature attributes

Graphic features of map layers and their data records are connected, so you can start with a feature and view its record. You can also find features on a map using feature attributes.

## Use the Identify tool

To display the data attributes of a map feature, you can click the feature with the Identify tool. This tool is the easiest way to learn something about a location on a map.

**1**  **Save your map document as** Tutorial1-5.mxd **to the Chapter1 folder.**

**2**  **Click Bookmarks > 48 Contiguous States.**

**3**  **On the Tools toolbar, click the Identify button** 🛈. **Click anywhere on the map.**

**4**  **In the Identify window, click the Identify from list > US States.**

**5**  **Click inside the state of Texas.** The state temporarily flashes and its attributes appear in the Identify window. Note one of the field values for the state, such as Hispanic population, from the 2010 US Census.

Next, you use the Identify tool's options to control which features it processes.

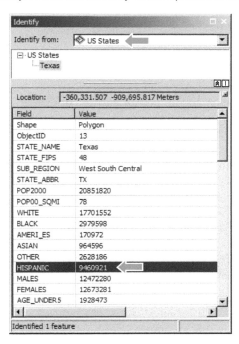

**6**   In the Identify window, click the Identify from drop-down list > US Cities.

**7**   Click the red circular point marker for Houston (at the southeastern side of Texas). Make sure the point of the arrow is inside the circle when you click the mouse button. Notice which feature flashes—that is the feature for which you get information. This table includes only basic population data for 2000 and 2007.

**8** Continue clicking a few other cities to see the Identify results. Press and hold the Shift key to retain information for more than one city. Then click the name of a city in the top panel of the Identify window to view that city's information.

## Use advanced Identify tool capabilities

You can use the Identify tool to navigate and create spatial bookmarks.

**1** Without pressing the Shift key, click Houston with the Identify tool.

**2** Right-click the name Houston in the Identify window and click Flash. This flashes Houston's point marker.

**3** Right-click the name Houston again and click Zoom To. The map display zooms to Houston, Texas. ArcMap identifies US Cities only because its layer is set in the window.

**4** Right-click the name Houston again and click Create Bookmark. A bookmark of this zoomed area, named Houston, will automatically be created.

**5** Close the Identify window.

**6** Click Bookmarks > 48 Contiguous States.

**7** Click Bookmarks > Houston.

> ## YOUR TURN
>
> Turn on layers for COCounties and COStreets. Restrict the Identify results to the COCounties layer and identify Colorado counties. Practice making bookmarks for various Colorado counties using the Identify tool. Close the Identify window.

## Find features

Use the Find tool to locate features in a layer or layers based on their attribute values. You can also use this tool to select, flash, zoom, bookmark, identify, or unselect the feature in question.

**1**  Click Bookmarks > 48 Contiguous States.

**2**  On the Tools toolbar, click the Find button 🏛 .

**3**  Click the Features tab.

**4**  Type Boston as the feature and click Find. The results appear in the bottom section of the Find window.

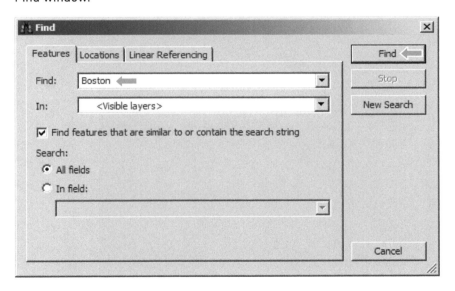

**5** Right-click the city name Boston.

**6** Click Zoom To. The extent zooms to the city of Boston, Massachusetts.

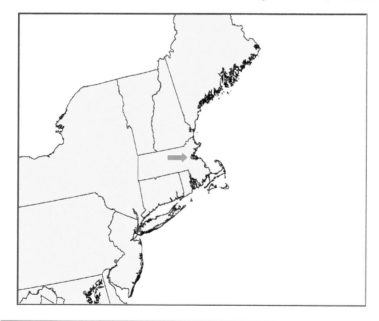

## YOUR TURN

Find other US cities and practice showing them using other find options such as Flash Features, Identify Feature(s), and Set Bookmark. When finished, clear any selected features, and zoom to the 48 Contiguous States bookmark. Save your map document.

## TUTORIAL 1-6
# Selecting features

You can work with a subset of one or more features in a map layer by selecting them. For example, before you move, delete, or copy a feature (as you will learn about in future chapters), you must select it. Selected features appear highlighted in the layer's attribute table and on the map.

## Use the Select Features tool

**1**   **Save your map document as** Tutorial1-6.mxd **to the Chapter1 folder.**

**2**   **Click Bookmarks > 48 Contiguous States.**

**3**   **Turn off the COStreets and COCounties layers.**

**4**   **On the Tools toolbar, click the Select Features button 🖯 ▾.**

**5**   **Click inside Texas.** This action selects Texas and highlights it with a blue outline.

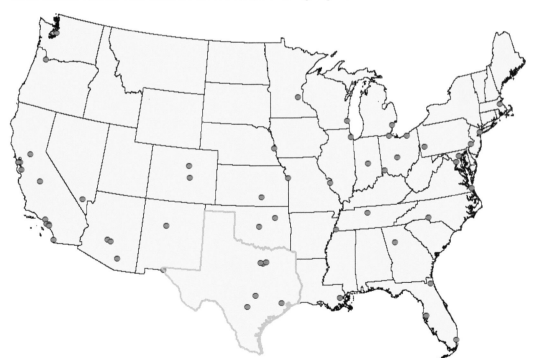

**6**  Press and hold the Shift key and click inside the four states adjacent to Texas.

**7**  Save your map document.

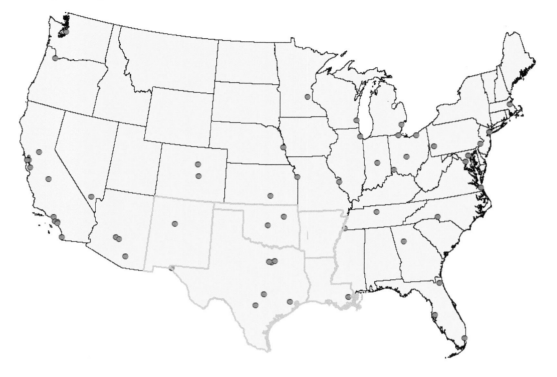

## TUTORIAL 1-7
# Changing selection options

Sometimes you will want to produce a map with certain features selected. Then it is desirable to be able to change the selection color for the purpose at hand.

**1**  Save your map document as Tutorial1-7.mxd to the Chapter1 folder.

**2**  On the Menu bar, click Selection > Selection Options.

**3** In the Selection Tools Settings frame, click the color box.

**4** Click Amethyst (second column from the right, row 3) as the new selection color and click OK.

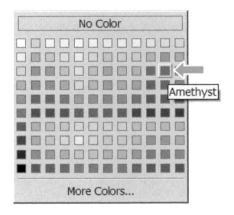

**5** On the Menu bar, click Selection > Zoom to Selected Features. The map will zoom to the selected features and the selection color for map features will now be amethyst.

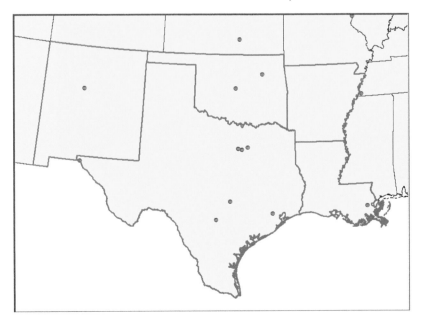

**6** On the Menu bar, click Selection > Clear Selected Features.

## Change selection properties by layer

In addition to changing the color of selected features, you can change the selection symbol for the entire map or for individual layers.

**1** In the Table Of Contents window, right-click US Cities, and click Properties. The resulting Layer Properties window is one that you will use often. It allows you to modify many properties of a map layer.

**2** Click the Selection tab and the symbol button.

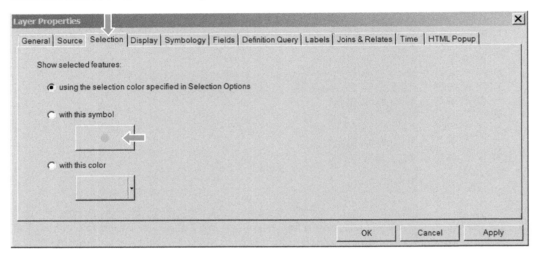

**3** Click Square 2, Amethyst color, and size 12, and click OK > OK.

**4** Click a city feature to see the new selection symbol, color, and size.

**5** Clear the selected features.

## YOUR TURN

Turn on and zoom to the COCounties and COStreets layers. Change the selection for Colorado counties (COCounties) to an Amethyst fill color with no outline. Select some counties to test the selection properties. When finished, turn off the COCounties layer. Change the selection color of the COStreets layer to Amethyst and select some streets.

## Set selectable layers

When there are many layers in a map document, you may want to restrict which ones are selectable. This simplifies the selection process and avoids selecting multiple features from different layers.

**1** Clear selected features and click Bookmarks > 48 Contiguous States.

**2** In the Table Of Contents window, turn off layers COCounties and COStreets.

**3**  **At the top of the Table Of Contents window, click the List By Selection button.**

**4**  **Clear the check boxes for COStreets, COCounties, and US States.** Now only US Cities will be selectable.

**5**  **Click the Select Features button and click a city.** The selected city gets the selection symbol and color that you chose previously and is listed in the table of contents.

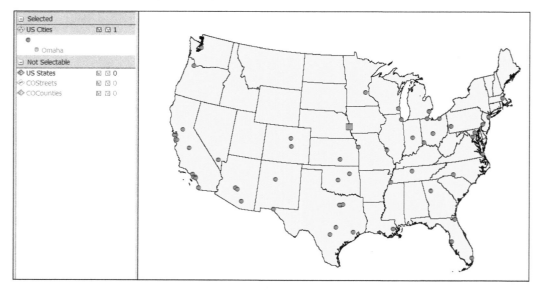

**6**  Clear the selected features.

# Select by graphic

Selecting features by using graphics is a shortcut to select multiple features.

**1**  Click Bookmarks > Florida.

**2**  Click the Select Features drop-down arrow and click Select by Circle.

**3**  Click inside the state of Florida and drag to draw a circle that includes the three cities in Florida.

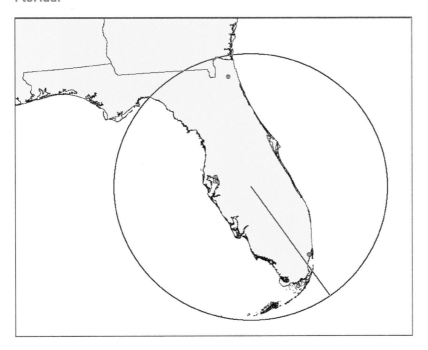

**4**   **Click Bookmarks > 48 Contiguous States.** The resulting map shows multiple cities selected and the resulting names in the table of contents.

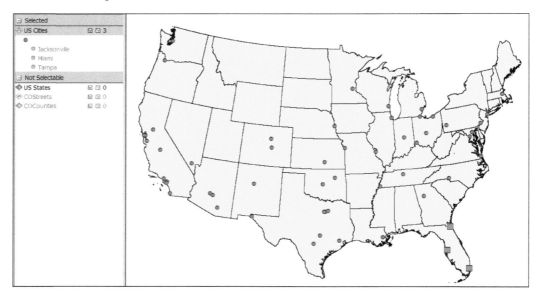

**5**   **Save your map document.**

## TUTORIAL 1-8
# Working with attribute tables

You can view and work with data for map features in the layer's attribute table.

## Open the attribute table of the US Cities layer and select a record

To explore the attributes of a layer on a map, open its attribute table and select a feature.

**1**   **Save your map document as** Tutorial1-8.mxd **to the Chapter1 folder.**

**2**   **In the Table Of Contents window, right-click US Cities and click Open Attribute Table.** The table opens, containing one record for each US City point feature. Every layer has an attribute table with one record per feature.

**3** Scroll down in the table until you find Chicago and click the record selector (gray cell on the left side of the row) for Chicago to select that record. If a feature is selected in the attribute table, it also is selected on the map.

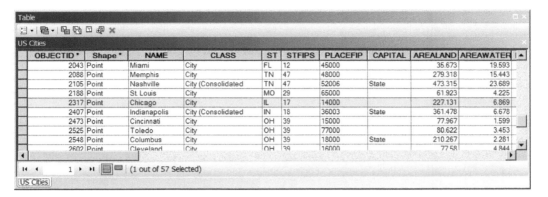

**4** Resize the US Cities table to see both the map and table on the screen. You should see the city of Chicago on the map.

**5** In the table, click the Clear Selection button 🔲.

## Select features on the map and see selected records

**1** Select all cities in Florida on the map.

**2** In the US Cities table, click the Show Selected Records button. This shows the records for only the features selected on the map: the cities in Florida.

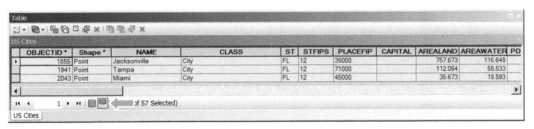

**3** Click the Show All Records button ▤ to show all records again. The Florida cities will still be selected.

## Switch selections

You can select records in a layer and reverse the selection.

**1** In the US Cities table, click the Table Options drop-down arrow ▤ ▾.

**2** **Click Switch Selection.** This reverses the selection and selects all records that were not selected (cities outside of Florida) and deselects those that were selected (Florida cities).

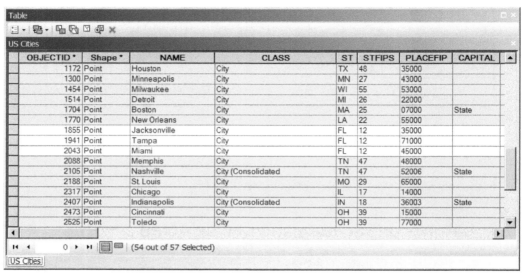

**3** Clear the selected features.

# Move a field

**1** In the US Cities table, click the gray title of the POP2000 field.

**2**  Drag the POP2000 field to the right of the NAME field.

| US Cities | | | | | | |
|---|---|---|---|---|---|---|
| OBJECTID * | Shape * | NAME | POP2000 | CLASS | ST | STFIP |
| 27 | Point | Honolulu | 371657 | Census Designated Place | HI | 15 |
| 88 | Point | Seattle | 563374 | City | WA | 53 |
| 130 | Point | Portland | 529121 | City | OR | 41 |
| 283 | Point | Oakland | 399484 | City | CA | 06 |
| 287 | Point | San Francisco | 776733 | City | CA | 06 |
| 325 | Point | San Jose | 894943 | City | CA | 06 |
| 377 | Point | Sacramento | 407018 | City | CA | 06 |
| 400 | Point | Fresno | 427652 | City | CA | 06 |
| 425 | Point | San Diego | 1223400 | City | CA | 06 |
| 469 | Point | Los Angeles | 3694820 | City | CA | 06 |
| 517 | Point | Long Beach | 461522 | City | CA | 06 |
| 560 | Point | Anaheim | 328014 | City | CA | 06 |
| 564 | Point | Santa Ana | 337977 | City | CA | 06 |
| 650 | Point | Las Vegas | 478434 | City | NV | 32 |
| 664 | Point | Tucson | 486699 | City | AZ | 04 |
| 696 | Point | Phoenix | 1321045 | City | AZ | 04 |

## Sort a field

**1**  **In the US Cities table, right-click the NAME field name and click Sort Ascending.** This sorts the table from A to Z by the name of each US city with a population over 300,000 in the year 2000.

| US Cities | | | | | | |
|---|---|---|---|---|---|---|
| OBJECTID * | Shape * | NAME | POP2000 | CLASS | ST | STFIP |
| 814 | Point | Albuquerque | 448607 | City | NM | 35 |
| 560 | Point | Anaheim | 328014 | City | CA | 06 |
| 879 | Point | Arlington | 332969 | City | TX | 48 |
| 2755 | Point | Atlanta | 416474 | City | GA | 13 |
| 1128 | Point | Austin | 656562 | City | TX | 48 |
| 3032 | Point | Baltimore | 651154 | City | MD | 24 |
| 1704 | Point | Boston | 589141 | City | MA | 25 |
| 2833 | Point | Charlotte | 540828 | City | NC | 37 |
| 2317 | Point | Chicago | 2896016 | City | IL | 17 |
| 2473 | Point | Cincinnati | 331285 | City | OH | 39 |
| 2602 | Point | Cleveland | 478403 | City | OH | 39 |
| 797 | Point | Colorado Springs | 360890 | City | CO | 08 |
| 2548 | Point | Columbus | 711470 | City | OH | 39 |
| 900 | Point | Dallas | 1188580 | City | TX | 48 |
| 790 | Point | Denver | 554636 | City | CO | 08 |
| 1514 | Point | Detroit | 951270 | City | MI | 26 |

**2**  **Right-click the POP2000 field and click Sort Descending.** This sorts the field from the highest-populated city to the lowest-populated city in year 2000.

---

## YOUR TURN

Move the POP2007 field to the right of POP2000. Sort this field in descending order.

# Use Advanced Sorting

**1** In the US Cities table, move the ST (State) field to the left of the POP2000 field.

**2** Right-click the ST field and click Advanced Sorting.

**3** Make selections as follows. Note: With cities first sorted by state (ST), there may be more than one city per state, so the secondary sort by POP2000 within state makes sense. However, a tertiary sort by POP2007 of records already sorted by state and then POP2000 will not make any changes in sorted records because it's very unlikely that two cities within a state will have the same POP2000 data values.

**4** Click OK. This sorts the table first by state and then by population of each US city for the year 2000 for that state.

| Shape * | NAME | ST | POP2000 | POP2007 | CLASS | STFIPS | |
|---------|------|-----|---------|---------|-------|--------|---|
| Point | Phoenix | AZ | 1321045 | 1502129 | City | 04 | |
| Point | Tucson | AZ | 486699 | 526777 | City | 04 | |
| Point | Mesa | AZ | 396375 | 463132 | City | 04 | |
| Point | Los Angeles | CA | 3694820 | 3908521 | City | 06 | |
| Point | San Diego | CA | 1223400 | 1301514 | City | 06 | |
| Point | San Jose | CA | 894943 | 946979 | City | 06 | |
| Point | San Francisco | CA | 776733 | 796417 | City | 06 | |
| Point | Long Beach | CA | 461522 | 486861 | City | 06 | |
| Point | Fresno | CA | 427652 | 475005 | City | 06 | |
| Point | Sacramento | CA | 407018 | 462910 | City | 06 | |
| Point | Oakland | CA | 399484 | 411240 | City | 06 | |
| Point | Santa Ana | CA | 337977 | 346599 | City | 06 | |
| Point | Anaheim | CA | 328014 | 348374 | City | 06 | |
| Point | Denver | CO | 554636 | 578062 | City | 08 | |
| Point | Colorado Springs | CO | 360890 | 397815 | City | 08 | |
| Point | Washington | DC | 572059 | 591318 | City | 11 | |

1

## YOUR TURN

Move and sort by other field names. Try sorting by other multiple fields. For example, you could sort US Cities alphabetically or by state capitals.

## Get statistics

You can get descriptive statistics, such as the mean and maximum value of an attribute, in ArcMap by opening a map layer's attribute table using the Statistics function.

**1**  Close the table and click Bookmarks > 48 Contiguous States.

**2**  In the Table Of Contents window, right-click US States, click Selection, and click Make This The Only Selectable Layer.

**3**  Press and hold the Shift key and use the Select Features tool to select the state of Texas and the four states adjacent to it.

**4**  In the Table Of Contents window, right-click US States and click Open Attribute Table.

**5**  Scroll to the right, right-click the column heading for the POP_2010 field, and click Statistics. The resulting window has statistics for the five selected states; for example, the mean 2010 population is 7,681,076.

## YOUR TURN

Get statistics for the same states in 2000. The mean population should be 6,652,779. Get population statistics for a new selection of states or another field of your choice. Fields in this table are from the 2010 Census. Clear all selections when finished.

# Turn fields off

Some attribute tables will have many fields and the attribute table can be confusing. You can easily turn fields on or off. If you export a table with a field off it will be excluded from the resulting saved table. In this exercise you will turn on only fields related to population and race.

**1**  In the Table Of Contents window, right-click US States and click Properties.

**2**  Click the Fields tab and the Turn all fields off button ⬚ .

**3**  Turn on the fields STATE_NAME, WHITE, BLACK, AMERI_ES, ASIAN, OTHER, HISPANIC, and POP_2010 and click OK.

**4**  In the Table Of Contents window, right-click US States and click Open Attribute Table. The table should display population and race data only for 2010.

**5**  Close the attribute table.

## YOUR TURN

Turn off all fields except population 2010 and race for COCounties. Save your map document.

## TUTORIAL 1-9
# Labeling features

Labels are text items on the map derived from one or more feature attributes. ArcMap places labels dynamically depending on map scale.

## Set label properties and label features

There are many label properties that you can set. Here you get started by setting the data value source.

**1**  **Save your map document as** Tutorial1-9.mxd **to the Chapter1 folder.**

**2**  **Close all tables and clear selected features.**

**3**  **Click Bookmarks > Florida.**

**4**  **In the Table Of Contents window, right-click US Cities and click Properties > Labels tab.**

**5**  **Click the Label Field arrow and click NAME if it is not already selected.**

**6**  **Click the Symbol button > Edit Symbol and click the Mask tab > Halo.**

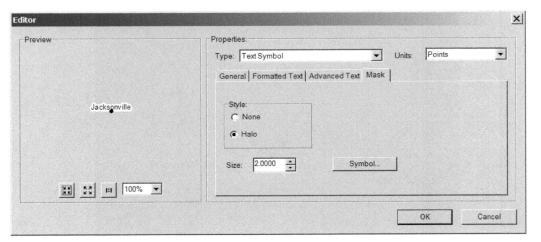

**7**  Click OK > OK.

**8**   Check the Label features in this layer check box and change the font to bold.

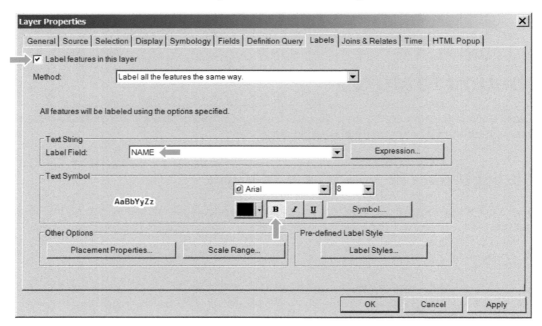

**9**   Click OK.

**10**  Zoom out to see additional states.

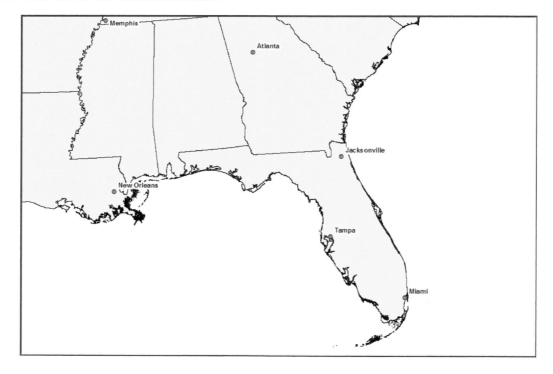

# Turn labels off

**1** In the Table Of Contents window, right-click US Cities and click Label Features.

**2** Click Label Features again to turn them back on.

# Convert labels to annotation

You can convert labels to graphics in order to edit them individually. You can convert all labels, only labels in a zoomed window, or labels from selected features only.

**1** Click Bookmarks > Florida.

**2** In the Table Of Contents window, right-click US Cities and click Convert Labels to Annotation.

**3** Make selections as shown in the figure, clicking In the map and Features in current extent, to label features in the state of Florida only.

**4** Click Convert.

# Edit a label graphic

Once labels become graphics, you can move, scale, and otherwise change them individually.

**1** On the Tools toolbar, click the Select Elements button ⬉.

**2** Click the text label for Miami and move it into the state of Florida.

**3** Similarly, move the label for Jacksonville.

**4** Save your map document and close ArcMap.

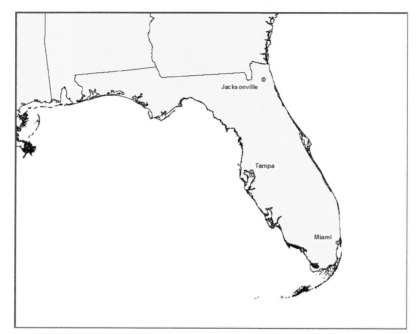

## ASSIGNMENT 1-1

# Analyze population by race in the top 10 US states

In this assignment, you will generate statistics for selected race fields for the 10 most populous US states and territories to generate a report with a map graphic and table generated from the GIS attribute table and Microsoft Excel. In chapter 4, you will learn how to calculate data directly in an ArcGIS table. In chapter 5, you will learn how to download census data as used in this assignment.

## Get set up

First, rename your assignment folder and create a map document.

Rename the folder **\EsriPress\GIST1\MyAssignments\Chapter1\Assignment1-1YourName\** to your name or student ID. Store all files that you produce for this assignment in this folder.

- Create a new map document named **Assignment1-1YourName.mxd** with relative paths.

You'll learn about map projections in chapter 5. Here, use the following steps to project the Table of Contents layer to a common projection for the 48 contiguous states:

- Right-click Layers in the table of contents and click Properties > Coordinate System tab.
- Expand Projected Coordinate Systems > Continental > North America and click North America Albers Equal Area Conic > OK.

## Build the map

Add the following to your map document:

- \EsriPress\GIST1\Data\UnitedStates.gdb\USStates—polygon features of US states and territories with Census 2010 data. An attribute is STATE_ABBR = two-letter state abbreviation.

## Requirements

- Symbolize the layer with a hollow fill and a medium-gray outline (30 percent).
- Zoom to the 48 contiguous states and create a spatial bookmark.
- Label states with their two-letter abbreviation using a yellow halo mask label. **_Hint:_** To create a halo label, right-click the layer in the table of contents and click Properties > Labels. Click

the Symbol button > Edit Symbol and click the Mask tab. Click Halo, click the Symbol button, and choose yellow as the symbol color.
- Change the selection color to Mars Red and use the USStates attribute table to select the 10 US states and territories with the highest population in 2010.
- Click File > Export Map and save your map as **Assignment1-1YourName.jpg** with 150 dpi resolution.

## Get statistics

Using the USStates attribute table, select the top 10 states and territories in 2010 by population to generate the first data column (Population by race: 10 most populous states and territories) in a Microsoft Excel spreadsheet similar to the following table. Use the same attribute table without selections to create the second data column (Total population by race: All states and territories). Finally, use an Excel expression to create the third data column (Percentage of population by race: 10 most populous states and territories). If you have trouble using Excel, you can create a table in Microsoft Word, manually enter the race data, and use a calculator for the percentage data.

Get the statistics from the ArcMap document. In the attribute table, select a statistic in the Statistics output table, press Ctrl+C to copy the statistic, click in the appropriate cell of your Excel or Word table, and press Ctrl+V to paste it.

In Excel, format cells for the first two data columns to use a thousands separator (,) and zero (0) decimal places. Use the expression = first data column/second data column *100 to get the third percentage column where you substitute appropriate cell addresses for the first and second columns.

|  | Population by race: 10 most populous states and territories | Population by race: All states and territories | Percentage of population by race: 10 most populous states and territories |
|---|---|---|---|
| White |  |  |  |
| Black |  |  |  |
| American Indian/ Inuit |  |  |  |
| Asian |  |  |  |
| Hawaiian/Pacific Islander |  |  |  |
| Other |  |  |  |
| Total: |  |  |  |

## Create a Word document

Create a Microsoft Word document named **Assignment1-1YourName.docx** and include the following:

- Title
- Your name
- Paragraph mentioning each exhibit that follows and any patterns that you observe

- Exported map inserted—include a caption at the top, such as **Exhibit 1. Top 10 most populous states and territories, year 2010**.
- Excel or Word table pasted—include a caption at the top such as **Exhibit 2. Statistics on populations by race**.

## ASSIGNMENT 1-2
# Produce a crime map

Crime prevention depends to a large extent on informal guardianship, meaning that neighborhood residents keep an eye on suspicious behavior and, if necessary, intervene in some fashion, such as calling the police. Neighborhood crime watch groups enhance guardianship, so police departments actively promote and support them and keep them informed on crime trends. Suppose that a police officer of a precinct has a notebook computer and projector for use at crime watch meetings. Your job is to get the officer ready for a meeting with the 100 Block Erin Street crime watch group of the Middle Hill neighborhood of Pittsburgh, Pennsylvania.

## Get set up

- Rename the folder **\EsriPress\GIST1\MyAssignments\Chapter1\Assignment1-2YourName\** to your name or student ID. Store all files that you produce for this assignment in this folder. Create a new map document named **Assignment1-2YourName.mxd** using relative paths.

## Build the map

Add the following to your map document:

- \EsriPress\GIST1\Data\Pittsburgh\Midhill.gdb\Bldgs—polygon features for buildings in the Middle Hill neighborhood of Pittsburgh.
- \EsriPress\GIST1\Data\Pittsburgh\Midhill.gdb\CADCalls—point features for 911 computer-aided dispatch police calls in the Middle Hill neighborhood of Pittsburgh. Attributes include NATURE_COD = call type, CALLDATE = date of incident, and ADDRESS = address of incident location.
- \EsriPress\GIST1\Data\Pittsburgh\Midhill.gdb\Curbs—Line features for curbs in the Middle Hill neighborhood of Pittsburgh.
- \EsriPress\GIST1\Data\Pittsburgh\Midhill.gdb\Streets—Line features for street centerlines in the Middle Hill neighborhood of Pittsburgh. Attributes include FNAME = street name, LEFTADD1 = beginning house number on the left side of the street, LEFTADD2 = ending house number on the left side of the street, RGTADD1 = beginning house number on the right side of the street, and RGTADD2 = ending house number on the right side of the street.

# Requirements

- Display buildings and curbs as medium-light gray (20 percent), CADCalls as bright-red circles, and streets as *no color* (this is a shortcut to use street names from this layer for curb labels). Label street names using the FNAME field. Create a spatial bookmark of the zoomed area of 100 Erin Street. The 100 block of Erin Street is the segment of Erin Street where addresses range from 100 to 199 and is perpendicular to Webster and Wylie streets.
- Set CADCalls as the only selectable layer and select the crimes in the blocks on either side of Erin Street between Webster, Wiley, Davenport, and Trent. Although it appears that there are only 6 incident points, there are actually 13 total because multiple incidents are at the same locations. Create a new layer of these crimes named **Erin Street Crimes** using a blue square. Clear your selected features.
- Click File > Export Map and save your map zoomed to Erin Street as **Assignment1-2YourName.jpg** with 150 dpi resolution.
- For attributes of the Erin Street Crimes layer, turn on the NATURE_COD, ADDRESS, and CALLDATE fields only. In the Attributes of Erin Street Crimes table, click the Table Options button and export and save the selected records to a dBASE (.dbf) file named **Assignment1-2YourName.dbf**. Open the dBASE file in Microsoft Excel. When opening the dBASE file in Excel, from the Files of Type menu, select All Files (*.*). This allows you to choose the dBASE file. Change the column headings to CRIME TYPE, ADDRESS, and DATE. Save your new spreadsheet as **Assignment1-2.xlsx.**

# Create a Microsoft PowerPoint document

Create a PowerPoint presentation named **Assignment1-2YourName.pptx** that includes the following:

- Title slide including your name
- Slide with map of Erin Street Crime Block Watch including the exported map zoomed to Erin Street
- Excel table listing the crime type, address, and date

# 2

# Map design

In this chapter, you'll learn all the steps necessary to symbolize common types of maps. Most analytical maps convey information with polygons or points using either qualitative or numeric attributes. Sometimes, but less often, lines are the primary interest. For polygons, symbolization mostly uses color fill and boundary lines. For points, it's the size, shape, color fill, and boundary lines of point markers. For lines, it's the type of line (for example, solid or dashed), width, and color. Symbolization is easy to implement because ArcMap uses attribute values to automate drawing; for example, it can draw all food pantry facilities in a city with a black square point marker of a certain size and all soup kitchen facilities with a black circle of a certain size by using an attribute with a type of facility code values.

Learning objectives

- Symbolize maps using qualitative attributes
- Symbolize maps using quantitative attributes
- Create custom numeric classes
- Create normalized and density maps
- Create dot density maps
- Create fishnet maps
- Create group layers

## TUTORIAL 2-1
# Creating point and polygon maps using qualitative attributes

Placing objects of all kinds into meaningful classes or categories is a major goal of science. Classification in tabular data is accomplished using attributes with codes that have mutually exclusive and exhaustive qualitative values. For example, a code for size could have the values low, medium, and high. Any instance of the features with this code is displayed in only one of the classes (the values are mutually exclusive) and there are no more size classes (the values are exhaustive). In this tutorial, you learn how to symbolize mapped features—points, lines, and polygons—by class membership as available in code attributes.

## Start ArcMap and open a map document

**1**   On the taskbar, click Start > ArcGIS > ArcMap 10.8.x.

**2**   Open Tutorial2-1.mxd from the Maps folder. ArcMap opens a map with no layers added. You add the needed layers next.

**3**   Save the map document to the Chapter2 folder of MyExercises.

## Add and display polygons using a single symbol

The first layer of your map is an outline of New York City neighborhoods used as a reference layer.

**1**   Click the Add Data button, browse through the Data folder to NYC.gdb, and click Neighborhoods > Add. ArcMap draws the neighborhoods for New York City as filled polygons.

**2**   In the Table Of Contents window, click the polygon symbol and select Hollow symbol, type an outline width of 1.15, and select an outline color of Black. Click OK.

**3**   In the Table Of Contents window, right-click the Neighborhoods layer, click Properties, and click the Labels tab.

**4**   Click the *Label features in this layer* check box. For Label Field, select Name, and select 10 as the Text Symbol size.

Next, you add a white halo to the label to make the labels stand out and be more legible.

**5**   Click the Symbol button > Edit Symbol button > Mask tab > Halo option button.

**6**  Type 1.5 **for Size and click OK > OK > OK.**

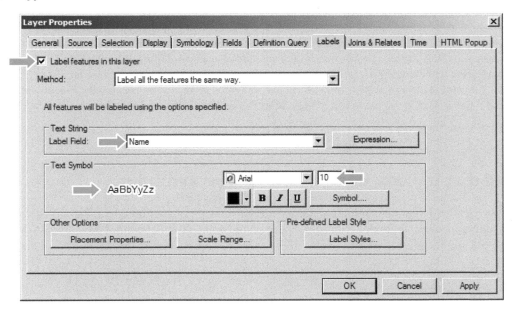

**7**  **Click Bookmarks > Lower Manhattan.** ArcMap zooms to this neighborhood and a few neighborhoods in the surrounding boroughs.

**Note:** The font size 10 labels for neighborhoods may appear larger or smaller than those shown on the book's map. If you'd like to change the size of labels for your window, use the steps in this section. Likewise, you can change the font size for labels in the Your Turn assignment, which is next.

**YOUR TURN**

Add the Water feature class from NYC.gdb with a blue symbol and no outline. Label the layer using the field LANDNAME with Times New Roman, Bold, and Italic font of size 12. Drag this layer below Neighborhoods. The two layers you added are from different sources, so the boundaries do not match perfectly. Note: When labeling the Water feature class, you'll see that some labels are repeated. You can get just one label to appear by clicking the Placement Properties button at the lower left of the Labels tab and selecting Remove duplicate labels.

## Add and display polygons using unique symbols

The next layer you add has zoning polygons with various classes of land use. Zoning restricts land use to approved kinds; for example, areas zoned as residential cannot have businesses located within them.

1   Click the Add Data button, browse through the Data folder to NYC.gdb, and click ZoningLandUse > Add.

2   After adding the ZoningLandUse layer, drag it below the Neighborhoods layer in the table of contents.

3   In the Table Of Contents window, right-click the ZoningLandUse layer and click Properties > Symbology tab.

4   In the Show panel, click Categories > Unique Values.

5   Under Value Field, click LANDUSE and click Add All Values.

ArcMap assigns colors unique to each land use. Next, you'll assign colors that are commonly used in zoning maps.

**6** Double-click the symbol next to Battery Park City, click Fill Color, and choose Apple Dust (column 6, row 7 of the color palette) for the color and Gray 30% as the outline color. Click OK.

**7** Assign a Gray 30% outline to the remaining symbols using the following colors (with column number, row number of the color palette included in parentheses). You can change all the outline colors by multiple selection (pressing and holding the Shift key), right-clicking, and choosing Properties for Selected Symbol(s).

- Commercial, Rose Quartz (2,1)
- Commercial Restricted, Medium Coral (2,8)
- Commercial Retail, Tulip Pink (2,9)
- Commercial Service, Rose Dust (2,7)
- Heavy Manufacturing, Blackberry (11,10)
- Light Manufacturing, Lepidolite Lilac (11,1)
- Medium Manufacturing, Lilac Dust (11,7)

- Park, Medium Apple (7,3)
- Residential, Yucca Yellow (5,1)
- Residential/Lt Mfg, Light Sienna (4,9)
- Waterfront, Blue Gray Dust (9,7)

**8** Click Apply > OK.

## Label zoning features

**1** In the Table Of Contents window, right-click the ZoningLandUse layer and click Properties > Labels tab.

**2** Click Label features in this layer, select ZONE as the Label field, select 6 as the Text Symbol size, select Gray 60% as the color, and click OK. The result is a subtle label for zoning details.

### YOUR TURN

ArcMap has predefined symbols for some zoning and land-use features. To view these, double-click the Park symbol > Style References > Civic > OK. Scroll through the Civic polygon symbols and choose Park & Open Space. When finished, deselect Civic under Style References.

## Add points and create a definition query

Often a map layer has more classes of features than you want to display. In this case, you can use a definition query to select the desired subset from the larger collection. Below, you add a new layer to your map document: Facilities, which has various government and nonprofit facilities that provide services. Only three classes out of over 100 are needed for the map. Classes have both a numeric code (Facility_T) and a corresponding text code (Facttype__1), and the three needed facility classes are 4901 = Soup Kitchen, 4902 = Food Pantry, and 4903 = Joint Soup Kitchen and Food Pantry. While in common language, you use *and* to connect members of a collection, such as the three food facility types, in a query you must use *or*.

**1**  Click the Add Data button, browse through the Data folder to NYC.gdb, and click Facilities > Add.

**2**  In the Table Of Contents window, right-click the Facilities layer and click Properties > Definition Query tab.

**3**  Click the Query Builder button.

**4**  In the Query Builder window, double-click FACILITY_T.

**5**  Click = as the logical operator.

**6**  Click Get Unique Values. The resulting list has all unique values in the FACILITY_T attribute.

**7** In the Unique Values list, double-click 4901.

**8** Click Or, double-click FACILITY_T, click = as the logical operator, and double-click 4902.

**9** Repeat step 8, except use 4903 instead of 4902. The completed query, FACILITY_T = 4901 OR FACILITY_T = 4902 OR FACILITY_T = 4903, yields a layer with only food type facilities. If your query has an error, edit it in the lower panel of the Query Builder, or click Clear and repeat steps 4 through 8.

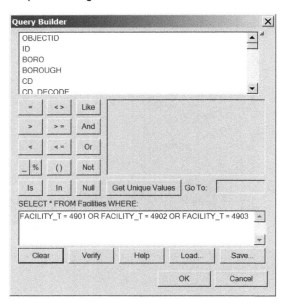

**10** Click OK > OK to execute your query and close the Layer Properties window.

## Display points using unique symbols

Next you symbolize food facilities using unique symbols.

**1** In the Table Of Contents window, right-click the Facilities layer and click Properties > Symbology tab.

**2** In the Show panel, click Categories > Unique Values.

**3** Click Factype__1 as the Value field and click Add All Values.

ArcMap assigns random unique symbols for each food facility type. Next, you assign three specific unique symbols.

**4** Double-click the symbol next to Food Pantry and select Esri symbol Square 1 and size 12. Click OK.

**5** Double-click the symbol next to Joint Soup Kitchen and Food Pantry and select Esri symbol Cross 2, size 12. Click OK.

**6** Double-click the symbol next to Soup Kitchen and select Esri symbol Circle 1, size 12. Click OK.

**7** Clear the check box next to <all other values> and click OK.

**8** Rename the Facilities layer Food Facilities and label using the field FACILITY_N, size 6 font, Fir Green text, and a white halo.

# Add and display lines using a single symbol

The final layer of your map is street centerlines, which is used as a reference layer. The street layer for all of New York City is very large, so to save space you add streets for Manhattan only.

**1**  Zoom to the Lower East Side neighborhood.

**2**  Click the Add Data button, browse through the Data folder to NYC.gdb, and click ManhattanStreets > Add.

**3**  In the Table Of Contents window, double-click the line symbol and select Gray 40%.

**4**  In the Table Of Contents window, right-click the ManhattanStreets layer and click Properties > Labels tab.

**5**  Click Label features in this layer and select Street as the Label field, 6 as the Text Symbol size, and Gray 70% as the color.

**6**  Click Placement Properties > Conflict Detection, select Low for Label Weight, and click OK > OK. **Note:** As a result of this step, other labels take precedence over streets and are displayed where space is limited.

**7** Save your map document.

## TUTORIAL 2-2

# Creating point and polygon maps using quantitative attributes

You cannot show continuous variation in points or polygons; the human eye simply cannot make distinctions unless there are relatively large changes in graphical elements. Generally, you must break up a numeric attribute into classes, similar to what you do to create a bar chart for a numeric attribute. Each bar has a minimum and maximum attribute value for its sides. Traditionally, the minimum value is included in the bar, but the maximum goes in the next bar to the right. All that is needed to symbolize map features is the set of maximum values for classes, which are called *break values*.

## Open a map document

**1** Open Tutorial2-2.mxd from the Maps folder. ArcMap opens a map with US Cities, US States, and US Counties layers added. You use these layers to create choropleth and graduated point maps.

**2** Save the map document as Tutorial2-2.mxd to the Chapter2 folder.

## Create a choropleth map of vacant housing units by county

A choropleth map uses color fill in polygons to represent numeric attribute values. Generally, increasing color value, or darkness of a color fill, in a color ramp represents either increasing or decreasing values. In this exercise, you will use US Census data to create choropleth maps for vacant housing units by county using the quantile classification method. Natural breaks (Jenks) is the default classification method that ArcMap uses, but you rarely use this method. Natural breaks uses an algorithm to cluster values of the numeric attribute into groups, with the boundaries of the groups' defining classes. You typically choose other methods for classifying your data, including your own custom classification, which you will do in tutorial 2-3.

**1** In the Table Of Contents window, right-click US Counties and click Properties > Symbology tab.

**2** In the Show panel, click Quantities > Graduated colors.

**3** Click VACANT for Value under Fields. This is the total number of vacant housing units by county from the 2010 census. The labels reflect the values in the VACANT column.

**4**  **Click the white-to-black color ramp.** This is the sixth color ramp from the bottom.

**5**  **Click the Classify button, choose Quantile for Classification, choose 5 as the number of classes, and click OK.**

**6**  **Click Label and click Format Labels.**

**7**  **Click Numeric under Category, type** 0 **for Number of decimal places in the Rounding section, and select the Show thousands separators check box.**

**8** Click OK > OK.

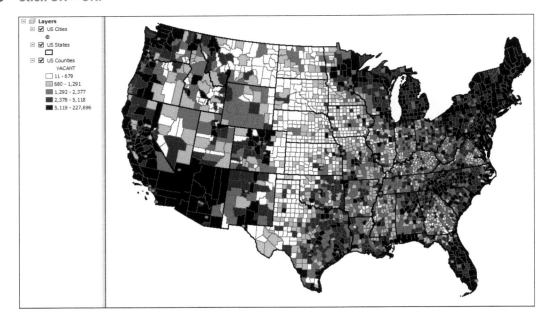

# Create a graduated point map of US cities by population

Point maps show locations of data or events using point markers. As a mathematical object, a point has no area, but to see a point on a map you must use a point marker with area, where the point is at the center of the marker. In the next exercise, you create a point map showing US major cities (2007 population over 250,000) as points with size-graduated point markers. The larger the marker, the higher the population.

**1** In the Table Of Contents window, right-click US Cities and click Properties > Symbology tab.

**2** In the Show panel, click Quantities > Graduated symbols.

**3** Click POP2007 for Value under Fields.

**4** Click the Symbol Selector button under Template, choose symbol Circle 2, choose Mars Red as the color, and click OK.

**5** Click the Classify button, choose Quantile for Classification, choose 5 as the number of classes, and click OK.

**6** Click the Label title, click Format labels, and click Numeric under Category.

**7** Type 0 for the Number of decimal places rounding and select the Show thousands separators check box. Click OK.

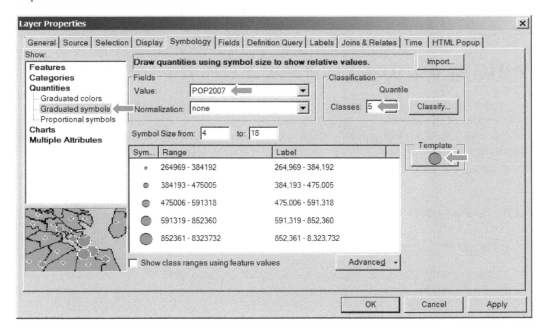

**8** Click OK. The map shows US Cities with low to high populations compared to US Counties with the highest number of vacant housing units. Do cities with high population appear to have the highest vacant housing units? A custom map will help answer this question.

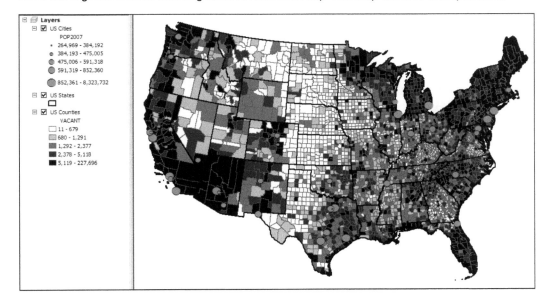

**9** Save your map document.

## TUTORIAL 2-3
# Creating custom classes for a map

You often need to create your own custom classifications. For example, if the frequency distribution of the attribute is long-tailed to the right (generally has low values but has a set of high values), then a good classification scheme has increasing width intervals in going from low to high values. This allows you to provide detailed information for the bulk of low values but also have classes for the high values.

## Create a custom classification

**1**   **Save a copy of your last map document as** Tutorial2-3.mxd **to the Chapter2 folder.**

**2**   **In the Table Of Contents window, right-click the US Counties layer and click Properties > General tab. Rename the layer** Vacant Housing by County.

**3**   **Click the Symbology tab and click the Classify button.**

**4**   **Click 6 for the number of classes and select Manual for classification method.**

Next, you create new breaks using increasing interval widths that double with each successive class.

**Note:** When typing break values for manual classification, as done in steps 5 and 6 below, always start by typing from the bottom up. If you type from the top down, ArcGIS sometimes does not allow you to enter needed values. Further note that you never have to change the maximum break value, so you start by typing the second-largest break value.

**5**   **In the Break Values panel, click the second-largest value, 6,045, to highlight it. Type** 96000 **and press Enter.**

**6** Continue clicking the break values, working your way up, and enter the following: 48000, 24000, 12000, **and** 6000.

**7** Click OK.

**8** In the Label field of the Symbology tab, change the first value to 6,000 or less **and the last label to** 96,001 or greater.

| Symbol | Range | Label |
|---|---|---|
| | 11.000000 - 6000.000000 | 6,000 or less |
| | 6000.000001 - 12000.000000 | 6,001 - 12,000 |
| | 12000.000001 - 24000.000000 | 12,001 - 24,000 |
| | 24000.000001 - 48000.000000 | 24,001 - 48,000 |
| | 48000.000001 - 96000.000000 | 48,001 - 96,000 |
| | 96000.000001 - 227696.000000 | 96,001 or greater |

**9** Click OK.

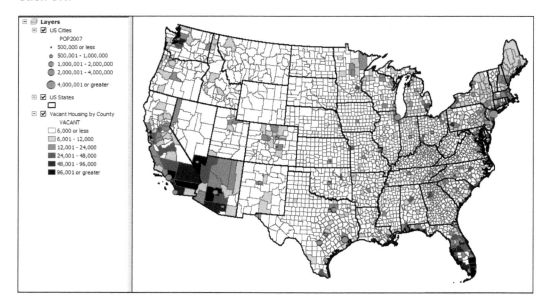

## YOUR TURN

Manually change the classes and labels for the 2007 population points to 500,000 or less, 500,001–1,000,000, 1,000,001–2,000,000, 2,000,001–4,000,000, and 4,000,001 or greater. Start by clicking value 852360 to change it to 4,000,000 first, and then change the values above it sequentially.

## Save a layer file

You can save symbolization of a layer as a layer file (.lyr) for reuse. This can save a lot of time and provide consistency between maps. For example, if you have to produce a new map every day to show recent events, it is efficient to apply a layer file to new data rather than go through all the interactive steps to symbolize from scratch each day.

**1** In the Table Of Contents window, right-click Vacant Housing by County, and click Save As Layer File.

**2** Browse to the Chapter2 folder of MyExercises, type Tutorial2-3 in the name field, and click Save.

# Import a layer file

Next, you will add the US Counties layer again and apply the custom classes that you saved in the layer file.

**1**  **Click the Add Data button, browse through the Data folder to UnitedStates.gdb, and click USCounties > Add.**

**2**  **In the Table Of Contents window, right-click the US Counties layer and click Properties > General tab.**

**3**  **Rename the layer** Owner Occupied by County**.**

**4**  **Click the Symbology tab > Import button.**

**5**  **Click the browse button, browse to your Chapter2 folder, and click Tutorial2-3.lyr > Add > OK.**

**6**  **Click OWNER_OCC as the Value Field and click OK > OK.**

**7**  **Drag the Owner Occupied by County layer to the bottom of the table of contents and turn the Vacant Housing by County layer off.**

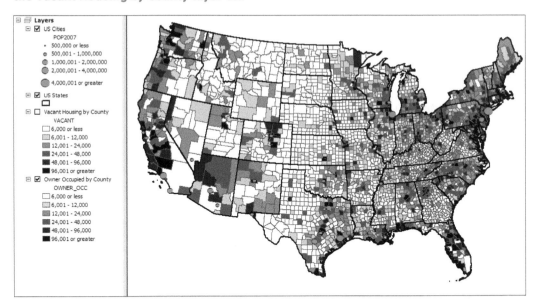

## YOUR TURN

Add the US Counties layer again and use Tutorial2-3.lyr to show renter-occupied housing units with the same classes as vacant and owner-occupied units. Rename the layer Renter Occupied by County and drag it to the bottom of the table of contents. When finished, save your map document.

## TUTORIAL 2-4
# Creating custom colors for a map

While ArcMap provides color ramps with preselected colors, you can change colors for classes manually. It is best to have more classes with light colors and a few with dark colors because the human eye can differentiate light colors more easily than dark ones. So here you create a custom color ramp that starts with white and ends with a dark blue.

## Manually change class colors

**1** Save your map document as Tutorial2-4.mxd **to the Chapter2 folder and turn off the Renter Occupied by County layer.**

**2** In the Table Of Contents window, right-click Owner Occupied by County, and click Properties > Symbology tab.

**3** Right-click the Color Ramp and click Properties.

**4** Click the color box next to Color 1 and click the Arctic White color if it is not already white.

**5** Select Color 2, click the color box next to Color 2, click the Dark Navy color, and click OK > OK. The Population by County map changes to reflect the new color ramp.

Next, you will change each color box manually.

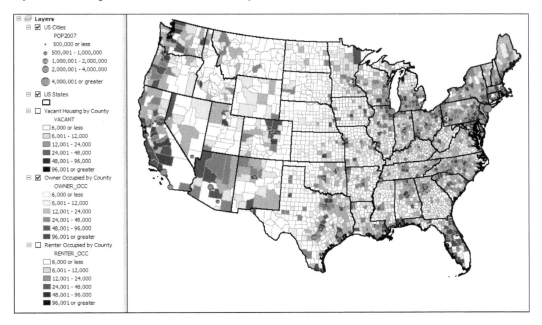

**6** In the Table Of Contents window, click the color box next to 6,001–12,000 for the Owner Occupied by County layer.

**7** Click the option next to Fill Color and click More Colors.

**8**  Type 5 for the Saturation (S) value and click OK > OK.

## YOUR TURN

Repeat steps 6 and 7 for the next two color boxes of the Owner Occupied by County layer and change the saturation to 20% and 40%. On the Symbology tab of the Layer Properties window, press Shift to select all classes. Click the Symbol title > Properties for Selected Symbol(s), and choose a Gray 30% outline to change the outline color for all color boxes of this layer. Save the layer file as Tutorial2-4.lyr and apply the layer with the new colors (and same classes) to the Renter Occupied by County layer. When finished, save your map document.

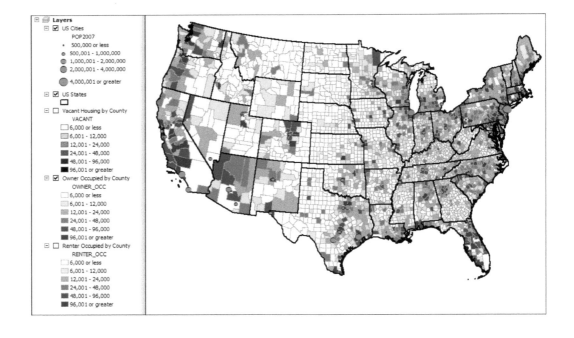

## TUTORIAL 2-5
# Creating normalized and density maps

If you symbolize a choropleth map with population data, you get information useful for doing things such as delivering services for the population. Services require capacities to match the size of the population; for example, a budget, materials, and labor to clean up vacant houses and lots in a city—cut high weeds, remove trash, eliminate vermin, and so forth. A different kind of information is available if you normalize population data. There are two options. One is on the relative sizes of population segments; for example, if you divide (normalize) the population of vacant houses by the total number of houses in a polygon, you get information on how well maintained or poor an area is. Areas with a low percentage of vacant houses are in good shape, while those with a high percentage are in bad shape. The second option is to divide population by the area of polygons, which yields density such as number of vacant houses per square mile. Density is a measure of congestion or concentration. A neighborhood with a high density of vacant houses is much worse off than another neighborhood with a low density. Many vacant houses near each other have a multiplier effect in dragging down a neighborhood.

## Create a choropleth map with normalized population

**1** **Save your map document as** Tutorial2-5.mxd **to the Chapter2 folder, turn on Vacant Housing by County, and remove Owner Occupied by County and Renter Occupied by County.**

**2** **In the Table Of Contents window, right-click the Vacant Housing by County layer and click Properties > General tab.**

**3** **Rename the layer** % Vacant Housing by County.

**4** **Click the Symbology tab and select HSE_UNITS as the Normalization field.** This displays the fraction of vacant housing units with VACANT as the numerator and HSE_UNITS as the denominator.

**5** **Click the Classify button, select 5 as the number of classes and Equal Interval for Classification, and click OK.**

**6** **Click the Label title > Format labels and click Percentage under Category.**

**7** Select the option for **The number represents a fraction. Adjust it to show a percentage.**

**8** Click the **Numeric Options button, type** 1 **as the number of decimal places, and click OK > OK. Click Apply to see your changes.** The following figure shows the final edits on the Symbology tab to show the percentage of vacant housing units.

**9**   **Click OK.** You now see the US counties with the highest percentage of vacant housing units, a very different map.

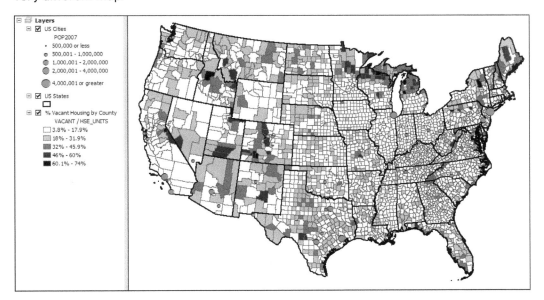

## Create a density map

The area field used in this exercise, ALAND10, was created by the US Census Bureau and has square meters for units. So the densities you display are vacant houses per square meter.

**1**   In the Table Of Contents window, right-click the % Vacant Housing by County layer and click Copy.

**2**   On the Menu bar, click Edit > Paste.

**3**   Turn off the original % Vacant Housing by County layer.

**4**  In the Table Of Contents window, right-click the copied layer, % Vacant Housing by County, and click Properties > General tab. Rename the layer Vacant Housing Density.

**5**  Click the Symbology tab, click ALAND10 for Normalization under Fields, and click OK.

**6**  In the Table Of Contents window, right-click US Cities > Label Features.

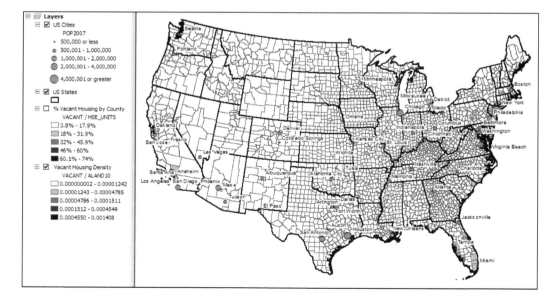

**7**  Click Bookmarks > New York. Notice the high density of vacant housing units.

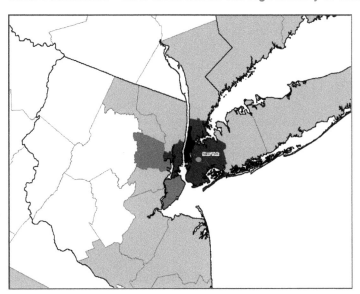

**8**  Click Bookmarks > Los Angeles. Notice that the density is not as high because the land area of Los Angeles County is much greater than the land area of New York County.

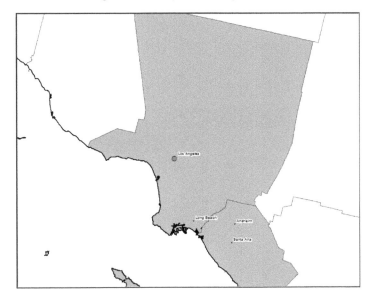

## YOUR TURN

Explore the cities under Bookmarks or other cities of interest to you. Click Bookmarks > 48 Contiguous States and save your map document.

## TUTORIAL 2-6
# Creating dot density maps

A dot density map is an alternative to the density map of the previous tutorial. Instead of color fill symbolizing polygons, each polygon is filled with randomly placed points based on the value of an attribute, generally a population. For example, each point might represent 100 people so that a polygon with a population of 1,000 would have 10 randomly placed points in it. The result is a type of area density map, where the percentage of area on a polygon covered with dots gives you an idea of how densely populated the polygon is.

## Open and save the map document

**1** **Open Tutorial2-6.mxd from the Maps folder.** ArcMap opens with a map of counties and states in the United States.

**2** **Save the map document as** Tutorial2-6.mxd **to the Chapter2 folder.**

**3** **Click Bookmarks > Pennsylvania.**

## Create a dot density map

Refer to the following figure for steps 3–6.

**1**   In the Table Of Contents window, right-click the US Population 2010 layer and click Properties > Symbology tab.

**2**   In the Show panel, click Quantities > Dot density.

**3**   Click POP_2010 as the Field Selection field and move the field to the right pane.

**4**   Double-click the dot symbol for POP_2010 and click Color > Mars Red > OK.

**5**   Click the line button in the Background section, select Gray 20% as the color, and click OK.

**6**   Type 5,000 for the Dot Value setting.

**7**   Click OK.

**8**   Save your map document.

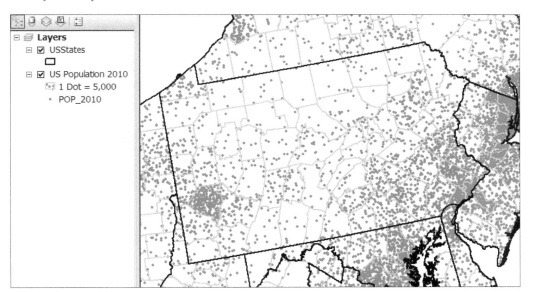

---

## YOUR TURN

Try changing the dot value to 10000 and 100000. Don't save the changes.

## TUTORIAL 2-7

# Creating fishnet maps

A fishnet map is so named because its uniform, square polygons (cells) make it look like a fishnet. Here you create a new polygon layer of uniform square cells and then aggregate point data (locations of public schools) to cell summaries (schools per cell). The advantage of a fishnet map is that when symbolized as a choropleth map, all polygon sizes and shapes are identical, so your perception of variation is driven only by color value. A choropleth map of irregular polygons, such as for counties, has the confusing nature of having two graphic elements that vary: color value and size/shape of polygon. Only the color value of polygons is valid: large-size, high-color-value polygons catch more of your attention, but polygon size has no direct meaning for the symbolized attribute.

## Open and save the map document

1   **Open Tutorial2-7.mxd from the Maps folder.** ArcMap opens with a map of counties and schools for Pennsylvania.

2   **Save the map document as** Tutorial2-7.mxd **to the Chapter2 folder.**

## Get set up for the fishnet tool

You must calculate the number of rows and columns for a given cell size to cover the extent of your map. The data folder has an Excel workbook that makes this easy. Cells must be square, and the map must have projected coordinates (not latitude and longitude). So cell size is simply the length of one side of a cell.

1   **Using File Explorer, browse to the Data > DataFiles folder and double-click FishnetCalculations.xlsx.** This worksheet has five inputs shown in a brown font. Cell size is the first, which is the length of a side of a cell. Next are four bounding coordinate values—for Pennsylvania, in this case. Top is the maximum y-coordinate, Bottom is the minimum y-coordinate, Right is the maximum x-coordinate, and Left is the minimum x-coordinate. The coordinates in the worksheet are correct for Pennsylvania. Cell size 30,000 is too large, however, so you will change it to 10,000. First, roughly check the coordinates.

| ▲ | A | B | C | D | E | F |
|---|---|---|---|---|---|---|
| 2 | | | | | | |
| 3 | Cell size | 30,000 | Can be changed to whatever you wish | | | |
| 4 | | | | | | |
| 5 | **Number of rows** | | | | | |
| 6 | Top | 4,726,414 | | | | |
| 7 | Bottom | 4,396,769 | | Divide Height by cell size | | |
| 8 | Height | 329,645 | | number cells = | 10.98818 | |
| 9 | | | | rounded up = | 11 | |
| 10 | **Number of columns** | | | | | |
| 11 | Right | 1,041,059 | | Divide Width by cell size | | |
| 12 | Left | 539,442 | | number cells = | 16.72056 | |
| 13 | Width | 501,617 | | rounded up = | 17 | |
| 14 | | | | | | |

2  **Using your cursor and coordinates readout at the lower right of the map window, verify the top, bottom, right, and left coordinates for PACounties.** While you may not get the identical coordinates in the workbook, you should be close. For a cell size of 30,000 meters (the units of this map), you would need 11 rows and 17 columns for the fishnet.

3  **Type** 10000 **for the cell size.** For this cell size, you need 33 rows and 51 columns.

4  **Save and close the Excel workbook.**

## Create the fishnet

1  **On the toolbar, click Windows > Search, type** fishnet **in the search textbox, press Enter, and click Create Fishnet (Data Management).**

**2** **Type or make selections as shown in the next figure.** The top, bottom, right, and left coordinates displayed in the Create Fishnet tool are alternative sources for these values instead of direct measurement on the map.

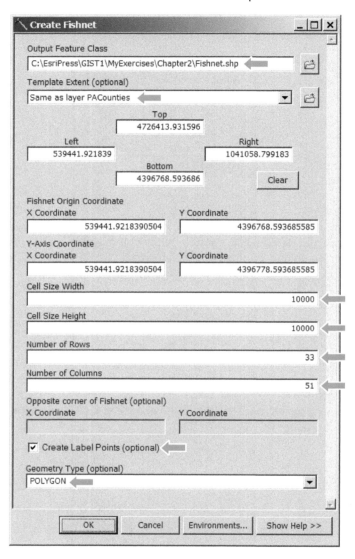

**3** Click OK and, if necessary, close the Search window.

**4** Turn off the Fishnet_label layer in the table of contents and symbolize Fishnet with a hollow fill and tan outline color of your choice. The Fishnet_label layer has centroid points for the cells.

**5** Move PACounties above the fishnet in the table of contents.

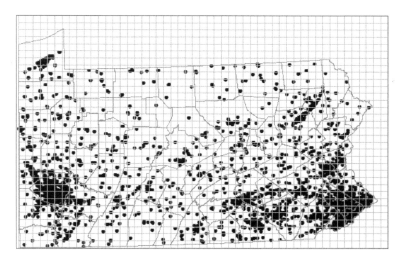

# Delete unneeded cells

Many cells lie outside Pennsylvania's boundary. Next, you eliminate all cells that are completely outside Pennsylvania.

1   Click Selection > Select By Location, and type or make selections as shown, with a source layer of PACounties, a target layer of Fishnet, and a spatial selection of intersect the source layer feature.

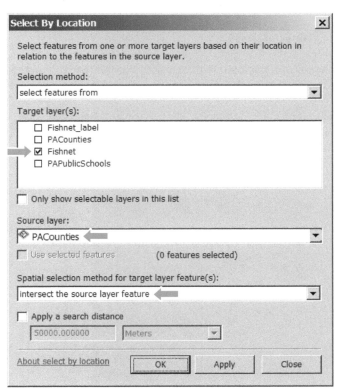

**2** Click OK.

**3** In the Table Of Contents window, right-click Fishnet, click Data > Export Data, and type or make selections as shown, with an export of Selected features, using the same coordinate system as this layer's source data, and an output feature class of FishnetPA.shp.

**4** Click OK and click Yes to add to the map.

**5** Right-click Fishnet_label and click Remove. Do the same for Fishnet.

## Spatially join schools to the fishnet

This step counts the number of schools per grid cell. With this step completed, you are ready to symbolize a choropleth map for number of schools per grid cell.

**1** In the Table Of Contents window, right-click FishnetPA, click Joins and Relates > Join, and type or make selections as shown, using Join data from another layer based on spatial location, the layer to join as PAPublicSchools, and the output shapefile as FishNetSchools.shp.

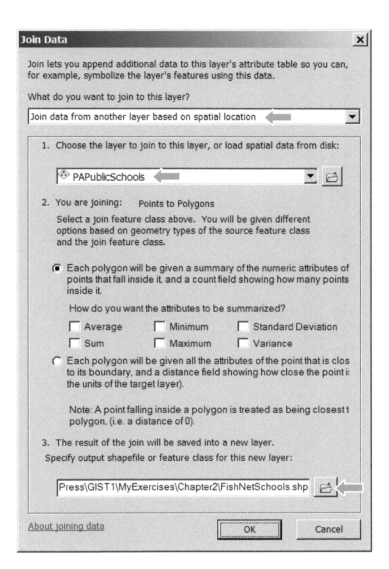

2  Click OK.

3  Remove FishnetPA and PAPublicSchools from the table of contents.

4  In the Table Of Contents window, right-click FishNetSchools and click Open Attribute Table.

5  Right-click the Count_ column heading and click Sort Descending. Count_ has the number of schools per grid cell, which has a maximum of 131 for grid cell 216.

6  Close the attribute table and move PACounties above FishNetSchools in the table of contents.

7  In the Table Of Contents window, right-click FishNetSchools and click Properties > Symbology.

**8**  Click Quantities and select Count_ as the value field.

**9**  Click Classify, select Quantile as the classification method, and click OK > OK.

**10**  Resymbolize PACounties to have a darker outline so that it shows up better. The uniform grid makes for a very clear choropleth map.

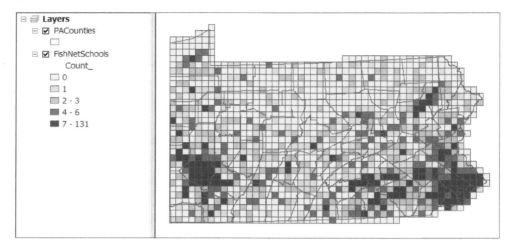

**11**  Select a suitable color ramp.

**12**  Save your map document.

## TUTORIAL 2-8
# Creating group layers and layer packages

If you have a map with many layers, some of which are in groups that would be convenient for the user to turn on and off as groups, then it is helpful to create group layers for that purpose. You can also save a group layer (or any single layer) as a layer package (a file with an .lpk extension). A layer package is one file with all data sources and symbolization included, which you can share with others, including online in ArcGIS Online. In this tutorial, you'll create layer groups for populations and facilities in New York City by administrative and political features including fire companies, police precincts, health areas, and school and council districts. You create layer packages from layer groups.

## Start ArcMap and begin a map document

**1**  Click File > New and click Blank Map > OK.

**2**  Save your map document as Tutorial2-8.mxd to the Chapter2 folder.

## Create a layer group and description

**1**  In the Table Of Contents window, right-click Layers and click New Group Layer.

**2**  Right-click the resulting new group layer and click Properties. The Group Layer Properties window appears.

**3**  Click the General tab and type NYC Fire as the layer name and NYC Fire Houses and Population as the description. A description is needed to later save the layer group as a layer package.

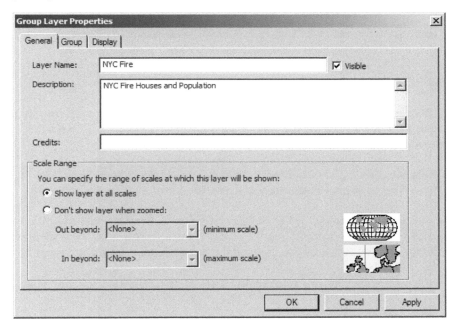

**4**  Click the Group tab > Add.

**5**  Browse through the Data folder to NYC.gdb; press and hold the Ctrl key; select Boroughs, Facilities, FireCompanies, and Water; and click Add (but do not click OK). These features are to be added to your NYC Fire layer group.

**6**  Click Boroughs and click the up arrow until Boroughs is at the top of the group layer list.

**7**  Click Water and click the down arrow so that Water is at the bottom of the group layer list.

**8** Click OK.

## Symbolize layers

**1** In the Table Of Contents window, right-click Boroughs; click Properties > Symbology tab; and choose Single symbol, hollow fill, and black outline with width 1.5. Leave the properties window open.

**2** Click the Labels tab. Check the Label features in this layer check box. Then click the Symbol button > Edit Symbol > Mask tab > Halo option button > OK > OK > OK.

**3** In the Table Of Contents window, right-click Facilities, click Properties > Definition Query tab, and type or select the definition query FACILITY_T = 2101. Click OK. This displays just the New York City fire houses. Now you can see the borough boundaries and labels.

**4** Rename the Facilities layer as Fire Houses and display as a Square 2 symbol, Mars Red, size 6.

**5** In the Table Of Contents window, right-click FireCompanies and click Properties > Symbology tab.

**6** Click Quantities > Graduated Colors, select Pop2010 as the Value field, select the black-to-white color ramp, click the Symbol heading above the color ramp chips, and click Flip Symbols (so that you get a white-to-black color ramp).

**7** Click the Classify button, select Quantile as the classification method, and click OK, leaving the Layer Properties window open.

**8** Click the Label heading over the right half of the symbols and click Format Labels > Show Thousands Separators > OK > OK.

**9** In the Table Of Contents window, right-click Water; click Properties > Symbology tab; choose Single symbol, Blue, with no outline; and click OK.

You now have a layer group for NYC Fire that includes four layers.

## YOUR TURN

Create a layer group named NYC Police with description NYC Police Precincts and Population. Add NYC.gdb layers Boroughs, Facilities, PolicePrecincts, and Water in the same order as the NYC Fire layers. Use the same symbology and labels as the NYC Fire layers. Use Pop2010 as the population field and FACILITY_T = 2001 (Police Stations). Symbolize Police Station points as Circle 2, the blue of your choice, and size 6. Rename this layer Police Stations. Try turning entire groups on and off by selecting the small check box to the left of each group layer name. Save your map document.

# Save a layer package

**1** In the Table Of Contents window, right-click NYC Fire and click Create Layer Package.

**2** In the Layer Package window, click the Save package to file option. ArcGIS should automatically fill in the package path and file name, such as C:\EsriPress\GIST1\MyExercises\ Chapter2\NYC Fire.lpk.

**3** Click the Item Description button and type NYC Fire Houses and Population in the Summary (required) field and NYC_Fire in the Tags (required) field.

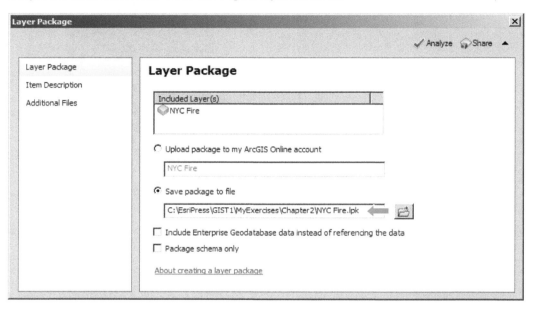

**4**   Click the Share button and wait while the layer package file is created.

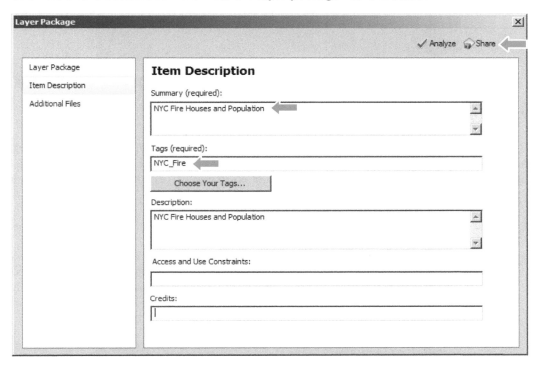

**5**   **Click OK when the Succeeded window appears.** You now have a layer package that can be shared with another ArcGIS user or uploaded to ArcGIS Online.

## YOUR TURN

Create a layer package for Police Precincts. When finished, save your map document and exit ArcMap.

## ASSIGNMENT 2-1

# Create a map showing schools in New York City by type

Suppose that a philanthropic foundation wants to study the role of charter schools in K–12 education in New York City, including accessibility of students to charter schools. An initial GIS project is simply to map all K–12 schools—including public, private, and charter schools.

## Get set up

- Rename the folder **\EsriPress\GIST1\MyAssignments\Chapter2\Assignment2-1YourName\** to your name or student ID. Store all files that you produce for this assignment in this folder.
- Create a new map document named **Assignment2-1YourName.mxd** with relative paths.

## Build the map

Create three group layers named **NYC Public Schools**, **NYC Public Charter Schools**, and **NYC Private Schools**. Include descriptions for each layer group.

Add the following to each layer group in the following order:

- \EsriPress\GIST1\Data\NYC.gdb\Boroughs—polygon features of New York City boroughs.
- \EsriPress\GIST1\Data\NYC.gdb\SchoolDistricts—polygon features of New York City school districts.
- \EsriPress\GIST1\Data\NYC.gdb\Facilities—point features of New York City facilities. Attributes include FACILITY_T = facility type and FacType___3 = Facility type descriptive name.
- \EsriPress\GIST1\Data\NYC.gdb\MetroRoads—line features of New York City major roads.
- \EsriPress\GIST1\Data\NYC.gdb\Water—polygon features of bodies of water in New York City.

## Requirements

- Symbolize Boroughs with a hollow fill, black, 2.0 outline width. Label the boroughs using a light-yellow halo mask, bold, size 12. Create spatial bookmarks for each borough.
- Symbolize School Districts with a hollow fill, ultra-blue, 1.75 outline width. Label the school districts using a white halo mask, bold, size 10, and field SchoolDist.
- Symbolize MetroRoads and Water as ground features and with labels.
- Create the following definition queries for each school type:
  - Public Schools FACILITY_T >= 1001 AND FACILITY_T <= 1006
  - Public Charter Schools FACILITY_T >= 1011 AND FACILITY_T <= 1017

- Private Schools FACILITY_T >= 1101 AND FACILITY_T <= 1106
- Symbolize each school type (public, public charter, private) using attribute FacTyp
  unique symbols for each type. For example, use circles for all public schools, sqι
  public charter schools, and triangles for all private schools. Use symbol Asterisk 2 ιυ.
  one special public charter school. Also, use different colors for each school type.
- Save the layer groups as layer packages named **NYCPublicSchools**, **NYCCharterSchools**, and
  **NYCPrivateSchools**.
- Save the map document with only the NYCPublicSchools layer group turned on and zoomed to
  Staten Island.

## ASSIGNMENT 2-2
# Create maps for military sites and congressional districts

Base Realignment and Closure (BRAC) is a process of the US federal government used by the
Department of Defense (DOD) and Congress to close and realign military installations to
reduce expenditures and achieve increased efficiency. At the same time, members of
Congress attempt to keep military installations open in their districts or states as sources of
employment and economic activity. GIS is an effective tool to visualize and analyze military
installations and congressional districts in the United States.

## Get set up

- Rename the folder **\EsriPress\GIST1\MyAssignments\Chapter2\Assignment2-2YourName\**
  to your name or student ID. Store all files that you produce for this assignment in this folder.
- Create a new map document named **Assignment2-2YourName.mxd** with relative paths.

Use the following steps to project the map to a common projection for the 48 contiguous states:

- Right-click Layers in the table of contents and click Properties > Coordinate System tab.
- Expand Projected Coordinate Systems > Continental > North America and click North America
  Albers Equal Area Conic > OK.

# Build the map

Create three group layers: (1) **111th Congress DoD Installation Count**, (2) **DoD Installation Type**, and (3) **BRAC Status**. Include descriptions for each layer group.

Add the following to each group layer according to the instructions and rename each layer appropriately. Include the US States layer in every layer group, symbolized to your liking and labeled using the state abbreviations:

- \EsriPress\GIST1\Data\UnitedStates.gdb\USStates—polygon features of US states.
- \EsriPress\GIST1\Data\UnitedStates.gdb\USCities—point features of US cities. An attribute is POP2007 = population of each city in the year 2007.
- \EsriPress\GIST1\Data\UnitedStates.gdb\Congress111—Polygon features of generalized boundaries for the 111th congressional districts. Attributes include DOD_COUNT = the number of DOD sites, installations, ranges, and training areas in each congressional district and PARTY (D = Democrat, R = Republican).
- \EsriPress\GIST1\Data\UnitedStates.gdb\MilitaryBnd—Polygon features of 2010 of DOD sites, installations, ranges, and training areas in the United States and Territories. An attribute is COMPONENT = type of military installation (Army, Navy, Air Force, Marine Corps, Washington Headquarters Service).
- \EsriPress\GIST1\Data\UnitedStates.gdb\MilitaryPt—point features of 2010 of DOD sites, installations, ranges, and training areas in the United States and Territories. An attribute is BRAC_SITE = site selected for closure under a BRAC action (YES or NO).

# Requirements

- Zoom to the 48 contiguous states and create a bookmark at this extent.
- In layer group 1, show the DOD count by congressional district as a choropleth map.
- In layer group 2, show and label US cities whose populations are greater than or equal to 250,000. Show DOD installations (MilitaryBnd polygons) as unique symbols using the COMPONENT field. Group the component colors based on the types. Use shades of purple for Air Force, green for Army, orange for Marine Corps, blue for Navy, and gray for Washington Headquarters Service. When choosing colors, focus on the active installations by making their colors darker than guard or reserve installations.
- In layer group 3, show DOD installations (MilitaryPt points) as unique symbols using the BRAC_SITE field and congressional districts as unique symbols using the PARTY field.

# Create maps for US veteran unemployment status

The US Census collects data about US veterans and their dependents. In this assignment, you create dot density and choropleth maps showing the unemployment status of veterans in the labor workforce by US county. The data for these maps consists of 2009 three-year estimates from the American Community Survey. Some counties are not populated enough and have no data. You will display these counties as a light-gray polygon.

## Get set up

- Rename the folder **\EsriPress\GIST1\MyAssignments\Chapter2\Assignment2-3YourName\** to your name or student ID. Store all files that you produce for this assignment in this folder.
- Create a new map document named **Assignment2-3YourName.mxd** with relative paths.

Use the following steps to project the table of contents layer to a common projection for the 48 contiguous states:

- Right-click Layers in the table of contents and click Properties > Coordinate System tab.
- Expand Projected Coordinate Systems > Continental > North America and click North America Albers Equal Area Conic > OK.

## Build the map

Add the following to your map document:

- \EsriPress\GIST1\Data\UnitedStates.gdb\USStates—polygon features of US states.
- \EsriPress\GIST1\Data\UnitedStates.gdb\USCounties—polygon features of US counties. Attributes include VET_LBRFORCE = number of US veterans in the labor force, VET_UNEMP = number of unemployed US veterans in the labor force.

## Requirements

- Zoom to the 48 contiguous states and create a spatial bookmark called **48 Contiguous States**.
- Symbolize USStates with a hollow fill, black, 1.5 point outline. Label every state with its abbreviation using a white halo mask, bold, size 8.

Add the USCounties layer three times, renamed, for the following maps:

- The first is a dot density map using field VET_LBRFORCE and a dark-blue dot representing 5,000 veterans in the workforce. Use a light-gray (20%) outline for the counties.
- The second is a choropleth map normalized showing the percentage of unemployed veterans in the workforce using the VETUNEMP/VET_LBRFORCE fields. Use a light-to-dark-red color ramp with five classes: 0–4%, 4.01–8%, 8.01–16%, 16.01–32%, and 32.01% and greater. Edit the label symbology format to show the numbers as percentages. Use the format The number represents a fraction with 2 decimal places.
- The third is a single-symbol map at the bottom of the table of contents. Use a light-gray (10%) fill color and a medium- to light-gray outline (30%). Because there are some counties with no veteran employment data, this is a way to show US counties with no data.

# 3

# GIS outputs

**ArcGIS can produce many forms of output, including interactive maps, printed maps for distribution, image files for use in presentations or on websites, map animations, and maps that you share using ArcGIS Online cloud services. In this chapter, you'll learn how to build each of these kinds of outputs.**

### Learning objectives

- Build an interactive GIS
- Build map layouts
- Build map animations
- Use ArcGIS Online

## TUTORIAL 3-1
# Building an interactive GIS

You already know from chapter 2 how to interact with a map document by zooming in and out, panning, using the Identify tool, and using other tools such as the Magnifier tool. A map document intended for interactive use, however, can have additional enhancements, and that's what this tutorial is about. For example, sometimes what you would like to see is highly detailed and cannot be viewed when at full extent, but must be viewed when zoomed in. ArcGIS provides visible scale ranges for map layers for this case, so depending on how far zoomed in or out you are, you see different layers. In addition, you need to learn more about labeling features; about a tool called map tips; and about turning map features, such as points, into hyperlinks to websites or files. Finally, to make data clear to the map user, you need to clean up attribute tables, hiding attributes of no use to the person viewing the map, and assigning self-documenting alias names to useful attributes. Note that you need an internet connection to display a satellite image that is included as a web service in the tutorial map document that you are about to open.

## Open a map document

**1**  **Start ArcMap and open Tutorial3-1.mxd from the Maps folder.** The map document opens with all layers visible at full extent. You can't make out Manhattan streets, which appear to be a brown-filled polygon instead of lines (only Manhattan streets are included instead of all New York City streets to save computer disk space). Also, the point features (food pantries and soup kitchens) are piled on top of each other so that you cannot select individual points for more information. Finally, the satellite image, while interesting to see, does not help with information on where to zoom next. Obscured are polygons for New York City's five boroughs, which, if visible and labeled, would be helpful for choosing areas to zoom in to.

**2**  **Save the map document to the Chapter3 folder of MyExercises.** Notice the scale of your map display on the Standard toolbar. The scale in the following image reads 1:500,000, but yours may differ depending on your computer screen size and ArcMap window size. The scale value means that one inch (or foot or any other unit for linear measurement) on the computer screen corresponds to 500,000 inches (or other unit) on the ground. The ratio 1:500,000 is small, so such a map scale is called *small scale*. You'll eventually zoom in to the West Side of Manhattan at a scale of approximately 1:30,000, which is a larger fraction and is called *large scale*. Obviously, the larger the scale, the more detail you can see and use.

# Set visible scale ranges

When you set a minimum scale range for a map layer (the minimum scale at which it displays), the layer does not display if you zoom out any farther than the minimum. When you set a maximum scale range (the maximum scale at which it displays), the layer does not display if you zoom in any farther than the maximum.

**1** **At the top of the Table of Contents window, click the List By Drawing Order button (the leftmost button).** This mode of the table of contents provides information on what layers are visible at the current map scale, after you assign visible scale ranges.

Next, you use bookmarks so that it is easy for you to follow instructions and assign visible scale ranges in this tutorial. The bookmarks are not an essential feature for the ultimate map user.

**2** **Click Bookmarks > Manhattan Borough. In the scale text box on the Standard toolbar, edit the denominator by adding** 1 **to it (so, for example, if it is 200,000, make it 200,001) and press the Tab key.** The last step, adding 1 to the scale denominator, is a shortcut so that when you use the Manhattan Borough bookmark in the future, the layers that you are about to modify display. You'll turn on the Satellite Image layer at this scale, making its current scale the minimum scale. Then at this scale or zoomed out farther, this layer will not display (but it will appear for the Manhattan bookmark scale, which is zoomed in one foot farther).

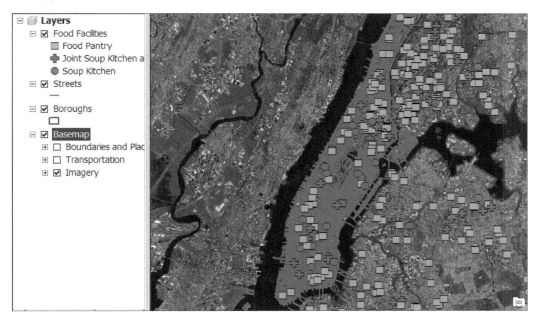

**3**   Right-click Basemap and click Visible Scale Range > Set Minimum Scale. Do the same for Food Facilities.

**4**   Click Bookmarks > NYC. Only Boroughs and Streets are on. Notice that the check marks for Food Facilities and Basemap are now shaded, signifying that these layers are available but not visible at the current scale.

**5**   Click Bookmarks > Manhattan Borough. All layers are on.

**6**   Click Bookmarks > Central Park West Side and add 1 to the denominator of the scale, similar to step 2.

**7**   Set the Minimum Scale Visible Range for Streets. Now try out all of the bookmarks, starting with NYC. NYC now displays just Boroughs. Manhattan Borough adds the Basemap image and Food Facilities. Central Park West Side adds Streets so that all layers are on.

## Label food facilities

While food facilities turn on at the scale of the Manhattan Borough bookmark, don't display labels for them until zoomed in farther to the Central Park West Side bookmark.

**1**   Click Bookmarks > Central Park West Side and add 1 to the denominator of the scale.

**2**   Right-click Food Facilities and click Properties > Labels. Select the Label features in this layer check box and verify that FACILITY_N is the label field.

First you will give the label a halo to make it highly visible. Then you will set its visible scale range.

**3**   Click the Symbol button > Edit Symbol > Mask > Halo.

**4**   Type 1.5 for size and click the Symbol button.

**5**   Select a pale green for fill color, No Color for outline color, and click OK > OK.

**6**   In the Symbol Selector window, set size to 6 and click OK.

**7**   In the Layer Properties window, click the Scale Range button > Don't show labels when zoomed, select <Use current scale> for Out beyond, and click OK > OK.

**8** Click Bookmarks > Central Park West Side. You should see labels for Food Facilities. They are very prominent because Food Facilities is the map layer of primary interest in this map document.

---

## YOUR TURN

Label Streets using Street as the label field with a size 6 font and 1.25 size halo that has a light-tan fill color and no outline. For Scale Range (Don't show labels when zoomed Out beyond), use the current scale. On the Labels tab of Properties, click Placement Properties > Placement > Parallel for orientation and On the line for position (turn off Above as a position).

Also, use BoroName as the label field for Boroughs with a size 8 font, 1.5 size halo, a light-blue fill for the halo, and no outline.

Finally, for the Central Park West Side bookmark, set a maximum scale for the Boroughs labels. First, add 1 to the scale denominator before setting the maximum scale. Then the Manhattan label should not display when at the Central Park West Side bookmark.

Try out all the bookmarks, starting with NYC. Notice that your street labels have priority over your more important Food Facilities labels: not all food facilities get labels. Next, you give Food Facilities labels the preference for display.

---

## Set priorities for labels

**1** Open the Labels tab of the Layer Properties for Food Facilities and click Placement Properties > Conflict Detection.

**2** Select High for both Label Weight and Feature Weight and click OK > OK.

**3** Open the Labels tab of the Layer Properties for Streets and click Placement Properties > Conflict Detection.

**4** Select Medium for Label Weight and Low for Feature Weight. Now labels for Food Facilities all display and take priority.

One more enhancement is to reduce the number of street labels created to cut down on clutter, using the next step.

**5** Type the value 1 in the Buffer field.

**6**  **Click OK > OK.** Now all the Food Facilities labels display, plus enough streets to get your bearings.

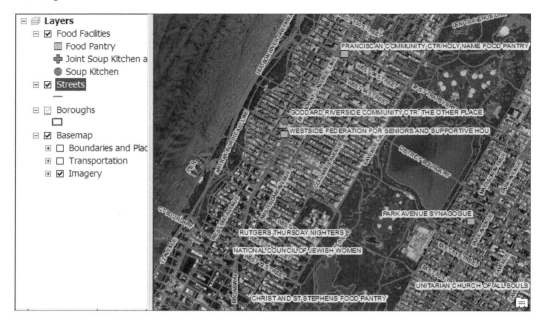

## Set field properties

When you use the Identify tool to get information, you should just see relevant attributes and they should have plain English names. Next, you set field properties of Food Facilities and Streets for this purpose.

**1**  **Open the Fields tab of the Layer Properties for Food Facilities and click the Turn Off All Fields button** ☷ **(to make them not visible). Select check boxes for FACILITY_N, FACILITY_A, ZIP, and Factype_1 to turn those fields back on (make them visible).**

**2**  **Click FACILITY_N to select it, and type** Name **as the Alias. Similarly, type** Address **as the alias for FACILITY_A,** ZIP Code **for ZIP, and** Type **for Factype_1. Click OK.** Now headers for these attributes will show the aliases instead of original names.

**3**  **Turn off all fields for Streets except Street and give it the alias** Name.

**4**  **Try using the Identify tool for Food Facilities and Streets to see the visible columns and aliases.** The results are useful and clearly labeled. This is very helpful to the map user.

## Set map tips

When you hover over a feature, you can get an attribute to display in a small window called a *map tip*.

1   Open the Display tab of Layer Properties for Food Facilities.

2   Select Address for the Display Expression and click Show Map Tips using the display expression. After you click OK, whenever you hover over a Food Facility, you will get its address in a map tip window.

3   Click OK and hover over some food facilities to see the map tips.

## Create map hyperlinks

If there were an attribute with URL website addresses for food facilities, you could select it on the Display tab of Properties and turn mapped features into hyperlinks. That's not the case here; nevertheless, you can manually enter some hyperlinks to features, which you do next.

1   Zoom in to the area with the food pantry, West Side Campaign Against Hunger.

2   Click the Identify tool on the Tools toolbar.

3   Click the point marker for the West Side Campaign Against Hunger food pantry.

4   In the Identify window, right-click the attribute value under Food Facilities in the top panel and click Add Hyperlink.

5   Click Link to URL, type http://www.wscah.org/ in the corresponding text box, click OK, and close the Identify tool.

6   Repeat steps 4 and 5 for the Goddard Riverside Community Ctr. and use http://goddard.org/ for the URL.

The remaining food facility in this group of three does not have a website yet. So instead you can place a hyperlink to a document—in this case, a PDF on your computer's hard drive.

7   With the Identify tool, click the point marker for the Westside Federation for Seniors and Supportive Housing, right-click its address, and click Add Hyperlink. Click the browse button for Link to Document, browse to the Data\DataFiles folder, and click Westside Federation.pdf > Open > OK. Close the Identify window.

**8**  Click the Hyperlink tool on the Tools toolbar 🗲 (all features with hyperlinks available get a blue dot in their point markers) and try out your hyperlinks. You must get the tip of the hyperlink tool's lightning bolt exactly in the center of a blue dot so that the lightning bolt turns black. Then you can click to activate the link.

**9**  Save your map document.

## TUTORIAL 3-2
# Creating map layouts

In this tutorial you learn how to create map layouts in ArcMap for use in a Word document, PowerPoint presentation, or website—or for distribution as paper maps. Map layouts have several elements, including a title, the map itself, a legend, a scale bar for ground distances, notes for data sources, and so forth. Rather than using prebuilt layout templates, you'll learn to build your own layouts from scratch—an essential skill.

## Open a map document and save a layer file

The set of layouts that you produce in this tutorial are for comparing populations by race or ethnicity for states. To facilitate comparisons of populations on separate maps, it's desirable to use the same numeric scale for all maps. Have you ever tried to compare bar charts in a newspaper or other publication where the horizontal-axis categories are the same but vertical axis scales are different? In your mind, you have to attempt to transform them to the same scale in some way to make comparisons. While you might be able to better represent each map separately with its own custom break points, for comparison purposes it's better to use the same break points for all maps. So as part of the work, you save a layer file that allows easy reuse of a numeric scale.

**1**  Open Tutorial3-2.mxd from the Maps folder.

**2**  Save the map document as Tutorial3-2.mxd to the Chapter3 folder.

**3**  Right-click Population in the Table of Contents window, click Save As Layer File, browse to the Chapter3 folder of MyExercises, type StatesPopulation.lyr for Name, and click Save.

The image shown is the custom layout you will build, shown in design mode where you construct layouts. The blue horizontal and vertical lines are guides that you create for precise alignment and placement of elements. They don't show up on the layout when you print it or save it as an image.

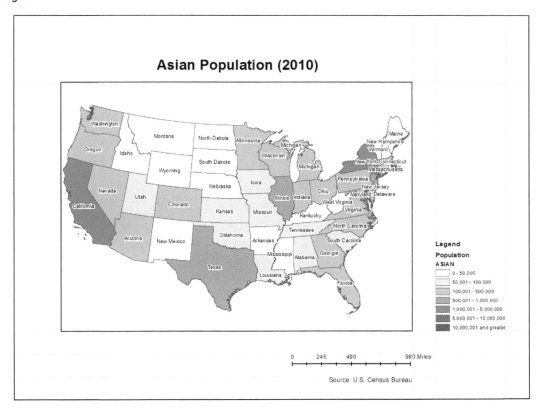

## Verify Layout View options

You need to verify option settings to work on layouts. You need settings to create and show guidelines and have layout elements snap to them. Snapping makes alignment of layout elements easy and precise.

**1**   On the Menu bar, click View > Layout View.

**2**   On the Menu bar, click Customize > ArcMap Options > Layout View tab.

**3**   Verify that your settings match those in the figure. Under Appearance, check Show
horizontal guides, Show vertical guides, and Show dashed line under active data frame.

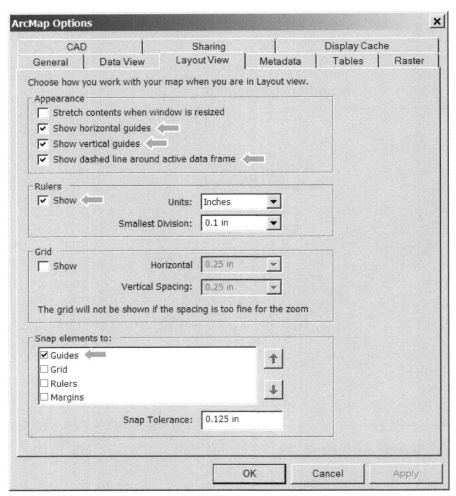

**4**   When finished, click OK.

## Set up layout page orientation and size

Next, set up the layout page using an 8.5-by-11-inch document.

**1**   Right-click anywhere in the white space outside the layout and click Page and Print Setup.

**2**   If you have access to a printer, select it and the desired properties, and then close any
windows for setting printer properties.

**3** In the ArcMap Page and Print Setup window, select Letter (8.5 by 11 inches) for paper size and Landscape for the Orientation in both the Paper and Page panels. Be sure to select the option buttons for landscape orientation in two places on the Page and Print Setup window.

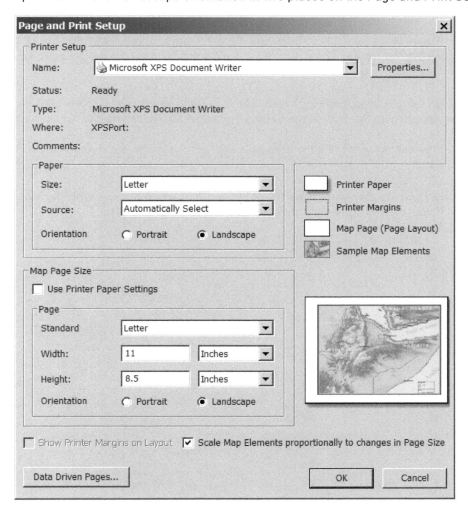

**4** Click OK.

**5** Right-click anywhere inside the layout and click the Zoom Whole Page button.

## Create and use guidelines in the layout view

In the next steps, you will use vertical and horizontal rulers to create guidelines.

**1** Click at 8.5 inches on the top horizontal ruler to create a vertical blue guide at that location. If you place your guide at the wrong location, right-click its arrow on the ruler, click Clear Guide, and start over. Alternatively, you can drag its arrow to relocate it.

**2**   Do the same at 7 inches on the left vertical ruler.

**3**   Click the map to select it (its frame outline and grab handles appear), right-click the map and click Properties, and then click the Size and Position tab.

**4**   Select the Preserve Aspect Ratio check box, type 7.5 in the Size Width field, press the Tab key, and click OK. Your map is now zoomed in too far, but its dimensions are good for the layout.

**5**   Drag the map so that its upper-right corner is at the intersection of the two guides. The map snaps precisely to the intersection of the guides when you release.

The objective of the next step is to fill the map element rectangle with the map as large as possible.

**6**   Use the Zoom tools to maximize the map of the contiguous 48 states within its map element rectangle. If you need to start over, click the Full Extent button.

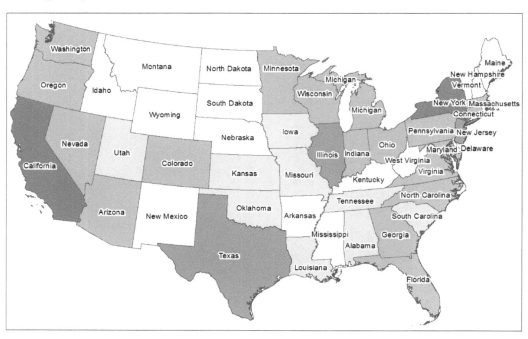

## Insert a title

**1**   On the Menu bar, click Insert > Title and double-click the title.

**2**   Type Asian Population (2010) in the text box, replacing the code enclosed in angle brackets.

**3**   Click Change Symbol, select 22 for size, and click B (Bold) for Style.

**4**   Click OK > OK.

**5** Center the title over the map.

## Insert a legend

**1** Click the horizontal ruler at 10.5 inches to create a new vertical guideline, and click the vertical ruler at 1.5 inches to create a new horizontal guideline.

**2** Click Insert > Legend (Population should be in the right panel, Legend Items), click Next four times, and click Finish.

Resizing and placing the legend takes the next three steps.

**3** Drag the legend so that its right side snaps to the 10.5-inch vertical guide and the bottom to the 1.5-inch horizontal guide.

**4** Click the horizontal ruler at 9 inches to create a new vertical ruler.

**5** Drag the upper-left grab handle of the legend to the right and down to make the legend smaller. Snap it to the 9-inch vertical guideline on the left while staying locked to the 10.5-inch vertical guideline on the right and 1.5-inch horizontal guideline on the bottom.

## Insert a scale bar

**1** Click the vertical ruler at 1 inch to create a new horizontal guide.

**2** Click Insert > Scale Bar.

**3** Click Scale Line 2 > Properties.

**4** Select Miles for the Division Units and click OK > OK.

**5** Drag the scale bar so that its top is at the 1-inch horizontal guideline and its right side is at the 8.5-inch vertical guideline.

**6** Right-click the scale bar and click Zoom to Selected Elements.

**7** Drag the left side of the scale bar to the right until its width is 1,000 miles. This takes trial and error, with you dragging and releasing to see the resultant width in miles.

**8** Right-click anywhere in the layout and click Zoom Whole Page.

## Insert text

**1** Click the vertical ruler at 0.5 inches to create a new horizontal guideline.

**2**  Click Insert, Text. ArcMap places a small text box in the center of your map (it is difficult to see at this scale).

**3**  Double-click the text box, type Source: US Census Bureau, and click OK.

**4**  Drag the text box so that its upper-right corner is at the intersection of the 8.5-inch vertical and 0.5-inch horizontal guides. This completes the layout.

**5**  Save your map document.

## Create a JPEG image of the layout

**1**  Click File > Export Map.

**2**  Save the image with file name Tutorial3-2 to the Chapter3 folder of MyExercises with Save as type JPEG and resolution 300 dpi (publication quality, a big file). Click Save.

**3**  Using File Explorer, browse to the Chapter3 folder of MyExercises and double-click Tutorial3-2Asians.jpg to open in a viewer. It is a sharp and professional-appearing map.

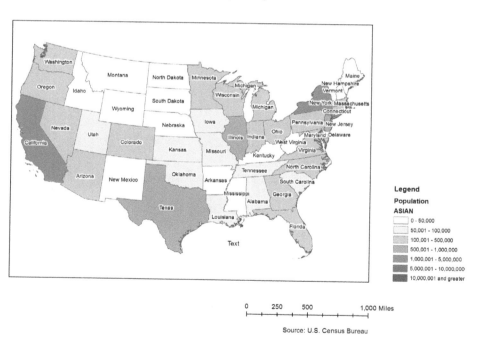

**4**  Close the viewer and File Explorer.

## TUTORIAL 3-3
# Reusing a custom map layout

Reusing your custom map to produce additional maps saves time, but just as important is the consistency of the resulting maps. Reuse guarantees that sizes and placement of objects match perfectly for a collection of maps.

## Open a map document

1  **Open Tutorial3-3.mxd from the Maps folder.** You will replace the Asian attribute with the Black attribute of the Population layer.

2  **Save the map document as** Tutorial3-3.mxd **to the Chapter3 folder.**

## Use a layer file

1  Click View > Data View.

2  Right-click the Population layer in the Table of Contents window and click Properties > Symbology tab > the Import button.

3  On the Import Symbology dialog box, click the browse button. Browse to the Chapter3 folder of MyExercises, double-click StatesPopulation.lyr, and click OK.

4  On the Import Symbology Matching dialog box, click the Value Field arrow and click Black > OK.

5  In the Symbology tab window, change the color ramp to a monochromatic blue ramp, right-click the color ramp (to the right of the label, Color Ramp), and click Properties > the arrow for Color 1 > the white color chip > button for Color 2 > its arrow > a dark blue > OK > OK.

6  Click View > Layout View.

**7**  **Change the map title to** Black Population (2010) **and click OK.**

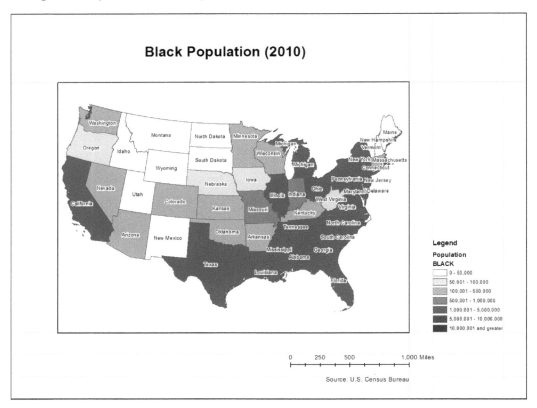

**8**  Save your map document.

## YOUR TURN

Make progress on completing the map collection of population by race by making a map document for one of the following populations from your template: White, Hispanic, or Native American. Name the map Tutorial3-3_YourTurn.mxd and save it to the Chapter3 folder of MyExercises. Use a monochromatic color ramp of your choice, but reuse the StatesPopulation.lyr layer file for break points.

## TUTORIAL 3-4

# Creating a custom map template with two maps

To facilitate comparisons, you can place two or more maps on the same layout. Your population maps by racial/ethnic groups are ideal for this purpose because they share the same numeric scale, making comparisons easy.

## Open a map document

**1** **Open Tutorial3-4.mxd from the Maps folder and save the map document as** Tutorial3-4.mxd **to the Chapter3 folder.**

**2** **Use your zoom tools to get the contiguous 48 states to fill the map window.**

**3** **Right-click the Population map layer and turn off Label Features for now.** Having to wait for ArcMap to add labels every time you make a change to the data frame takes too long when you get to layout view. So, to speed up map drawing, turn labeling off. You can turn it back on after the layout is finished. Nevertheless, ArcMap may slow down in reacting to your changes in this tutorial, so be patient. Layout view is a heavy user of computer resources.

## Create a second data frame

So far, you have worked with map documents that have only one data frame in the table of contents. It's possible to have two or more data frames, and this is necessary if you want to have two or more maps in a layout, as in this tutorial. While you could create a new data frame from scratch by clicking Insert > Data Frame, the easiest thing to do is to copy and paste the existing data frame, and then modify or add to its map layers. One of the benefits of copying and pasting is that the existing data frame has a map projection applied (Albers Equal Area Conic in this case), and when you copy and paste it to create a new data frame, the projection is retained. Otherwise, if you were to create a new data frame from scratch, you'd have to add the projection as a separate series of steps.

**1** **In the Table of Contents window, right-click Layers, click Properties > General tab, and change the name from Layers to** AsianPopulation **(but do not click OK).**

**2** **Click the Coordinate System tab and notice that the projection is USA Contiguous Albers Equal Area Conic.** This projection flattens the portion of the globe that includes the contiguous 48 states. Note under the projection name, there is a WKID. This is the Well-known ID, a code that indicates a specific projection.

**3**   Click OK > Yes.

**4**   Click Insert > Data Frame. Right-click the Population layer from the AsianPopulation data frame and click Copy.

**5**   Right-click the New Data Frame and click Paste. ArcMap adds a copy of the data frame to the table of contents.

**6**   Right-click the New Data Frame and click Properties > General tab. Rename the pasted data frame's name BlackPopulation.

**7**   Click the Coordinate System tab, and in the blank field, type 102003 and press Enter. This applies the Well-known ID of the USA Contiguous Albers Equal Area Conic projection to match the other data frame.

**8**   Click Yes to clear any transformation warning.

## YOUR TURN

For symbology of the Population map layer in the BlackPopulation data frame, right-click Population, click Properties, and click the Symbology tab. Import the StatesPopulation.lyr layer file using Black as the value field. Keep the same color ramp as the AsianPopulation data frame. Both data frames then can use the same legend in the map layout. Your finished map document will appear as shown. Note that to switch from one data frame to another, you right-click the data frame and click Activate. The active data frame's label in the table of contents is in bold type.

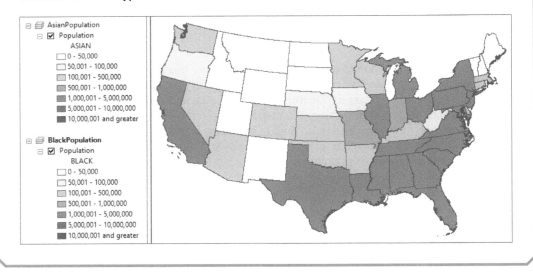

# Create guidelines in Layout View

Included is the layout that you will create in the steps of this exercise. It makes comparisons between two population distributions easy.

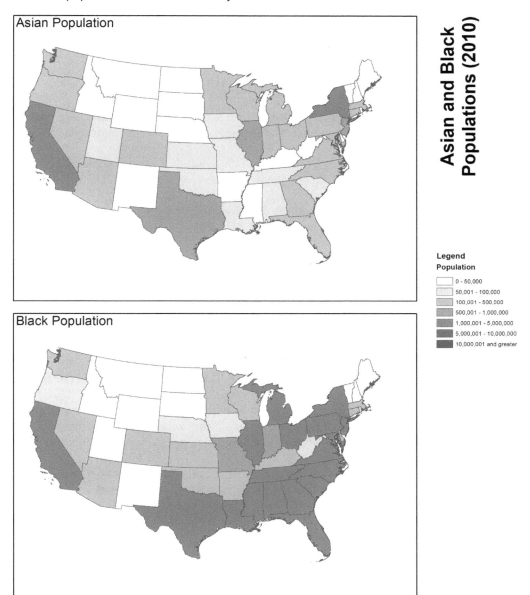

1  **Click View > Layout View.** Both data frames are in the view but may be directly over each other so that you see only one.

2  **Click the visible view, drag it up an inch or so, and release so that you see both views.**

**3** In layout view, right-click in the layout, click Page and Print Setup, make sure Size is Letter, make sure that both Portrait options are selected in the Page and Paper frames, and then click OK.

**4** Click the horizontal ruler at the 0.5-, 6.5-, 6.8-, and 8.0-inch marks.

**5** Click the vertical ruler at the 1.0-, 5.4-, 5.6-, 6.5-, and 10.0-inch marks.

## Relocate and resize data frames

**1** Make sure that the AsianPopulation data frame is active. Its data frame name needs to be bold in the table of contents and have grab handles visible on its boundary in the layout.

**2** Drag the AsianPopulation data frame so that its upper-left corner snaps to the intersection of the 10.0-inch horizontal guide and the 0.5-inch vertical guide.

**3** Drag the lower-right grab handle of the data frame to snap it at the 6.5-inch vertical guide. Drag a bottom-line grab handle to move the bottom of the data frame to the 5.6-inch horizontal guideline.

**4** Drag the BlackPopulation data frame so that its upper-left corner snaps at the intersection of the horizontal 5.4-inch guide and the vertical 0.5-inch guide.

**5** Drag the BlackPopulation data frame so that its lower-right grab handle snaps at the 6.5-inch vertical guide. Drag the bottom of the data frame to the 1.0 horizontal guide.

## Change and coordinate map extents of both data frames

The objective of this exercise is to get the map of each data frame to fill its frame and be identical in extent in both frames. Unfortunately, there is not a direct way to accomplish this, but there is a good guide available called *extent indicators* that you use to get close to the objective.

**1** Make sure that the AsianPopulation data frame is active. Use your Zoom and Pan tools to get the contiguous 48 states to fill up the data frame and be centered. Be careful that the Pan button does not get clicked on or remain on after use. It's easy to accidentally change the map view and extent with this tool. So always click the Select Elements tool on the Tools toolbar after using the Pan tool.

**2** Make the BlackPopulation data frame active.

**3** Right-click and click Properties > Extent Indicators.

**4** Move AsianPopulation from the left panel to the right panel and click OK.

**5** Click the Fixed Zoom Out button 🔲 on the Standard toolbar so that you see the entire red extent indicators.

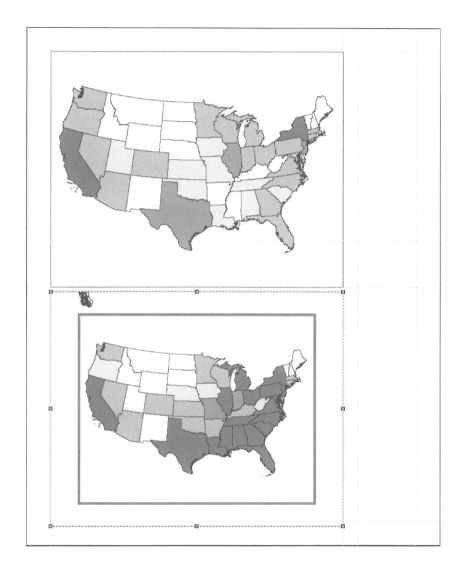

**6**   Use the Zoom In tool to drag a rectangle matching the red rectangle of the extent
indicators. You can repeat steps 4 and 5 until you get the BlackPopulation data frame close to
the extent of the AsianPopulation data frame.

**7**   Click the Select Elements button on the Tools toolbar. Right-click the BlackPopulation data
frame and click Properties > Extent Indicators. Move AsianPopulation from the right panel
to the left panel and click OK. Now both frames should have maps that fill up frames and
match well.

## Insert a legend

**1**   Click the AsianPopulation map element in the layout to activate its frame.

**2** Click Insert > Legend.

**3** Click Next four times, and then click Finish.

**4** Drag the legend so that it snaps on the upper right to the 8-inch vertical guide and 6.5-inch horizontal guide intersection, and then resize it to fit between the 6.8-inch and 8-inch vertical guides.

## Convert the legend to a graphic and modify it

The legend has the heading word "ASIAN," which you need to delete so that the legend is general and applies to both data frames. You can edit the legend by turning it into a graphic and ungrouping its parts.

**1** Right-click the legend and click Zoom to Selected Elements.

**2** Right-click the legend and click Convert To Graphics. Now the legend is no longer a "live object" that is automatically updated if you resymbolize the Asian Population layer. It's a static graphic, but now you can edit it.

**3** Right-click the legend and click Ungroup. While not needed here, you could further ungroup paint chips from their numeric intervals and edit the interval text.

**4** Click anywhere outside the legend to deselect its parts, right-click the ASIAN text box, and press Delete. Drag the Legend and Population text boxes down to place them closer to the rest of the legend. An alternative that often works better than dragging is to select an element, such as the Population text box, and use your arrow keys (down arrow in this case) to move it.

**5** Drag a rectangle around all legend parts to select them, right-click the legend, and click Group. Now the parts won't get separated.

**6** Right-click anywhere in the layout and click Zoom Whole Page.

## Insert a title

**1** Click Insert > Title. Click anywhere outside the title's text box, right-click the text box, and click Properties and the Text tab.

**2** In the Text panel, type Asian and Black, press Enter to jump to a new line, type Populations (2010), type 90 for Angle, and press Tab. Click Change Symbol, change the Size to 20, click B (bold) for style, and click OK > OK.

**3** Position the top left of the title text box at the intersection of the 6.8-inch vertical and 10-inch horizontal guides.

**4**  Save your map document.

---

**YOUR TURN**

Add a text box with the words Asian Population to the top map and another text box with Black Population to the bottom map. Export your layout to the Chapter3 folder of MyExercises as a JPEG image and view it in an image viewer. It is quite a nice layout and image.

---

## TUTORIAL 3-5
# Adding a report to a layout

Sometimes it's helpful to add a tabular report to a layout. The easiest thing to do is to export the map data of interest to Excel, create the tabular display you want, and copy and paste it to your layout. Here you use an Excel worksheet that is already prepared with a table of the top 10 states by Asian population.

## Open a map document

**1**  Open Tutorial3-5.mxd from the Maps folder and save the map document as Tutorial3-5.mxd to the Chapter3 folder. This is already a completed layout, except there is no data table.

## Open an Excel workbook

**1**  Launch Microsoft Excel, click File > Open, browse to the Data > DataFiles folder, and open AsianTop10States.xlsx. The worksheet has a nice table created from the population map layer's attribute table. The table was exported when it was open in ArcMap using the Table Options button and extra columns and rows were deleted, leaving the state names and populations of the 10 states with the highest Asian population. Features of the worksheet include that the grid is turned off (View tab, Show panel, Gridlines unchecked), borders are included for data cells (Home tab, Font panel, All Borders), and cells for the range A1:B12 (columns A and B and rows 1 through 12) have a white fill color (Home tab, Font panel, Fill Color, white).

| | A | B |
|---|---|---|
| 1 | Top 10 States | |
| 2 | State | Asian Population |
| 3 | California | 4,861,007 |
| 4 | New York | 1,420,244 |
| 5 | Texas | 964,596 |
| 6 | New Jersey | 725,726 |
| 7 | Illinois | 586,934 |
| 8 | Hawaii | 525,078 |
| 9 | Washington | 481,067 |
| 10 | Florida | 454,821 |
| 11 | Virginia | 439,890 |
| 12 | Massachusetts | 349,768 |

2    **Select the range A1:C13.** You must include an extra column and row in order to get all borders to copy. The extra column (C) and extra row (13) do not have color fill, so they will be invisible when pasted into the layout.

3    **Press Ctrl+C to copy the range.**

4    **Click the layout in ArcMap.**

5    **Click Edit > Paste.**

6    **Relocate and resize the table as shown on the map.**

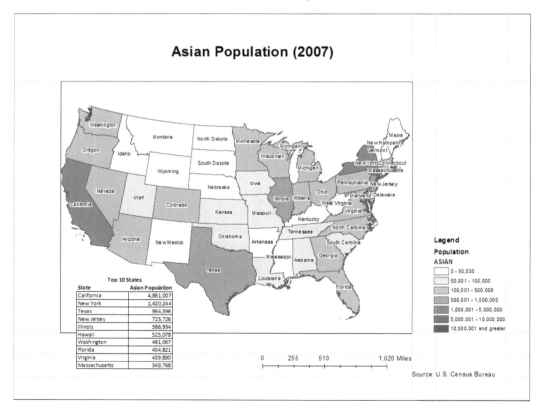

**7**   Save your map document.

## TUTORIAL 3-6
# Adding a graph to a layout

ArcMap has a convenient and easy means to create and add graphs to a layout.

## Open a map document

**1**   Open Tutorial3-6.mxd from the Maps folder. This is the same data view with which you started the previous tutorial.

**2**   Save the map document as Tutorial3-6.mxd to the Chapter3 folder.

## Sort and select records

**1**   Right-click Population in the Table of Contents window and click Open Attribute Table.

**2**   Right-click the ASIAN header and click Sort Descending.

**3**   Select the first 10 records (California through Massachusetts) by clicking and dragging the row selectors for those rows.

**4**   Close the attribute table.

## Create a graph and add it to a layout

**1**   Click View > Graphs > Create Graph. Then make selections as shown in the following image, with Horizontal Bar for Graph Type, Population for Layer/Table, ASIAN for Value field, ASIAN and Ascending for Y field, STATE_NAME for Y label field, and Add to legend unchecked.

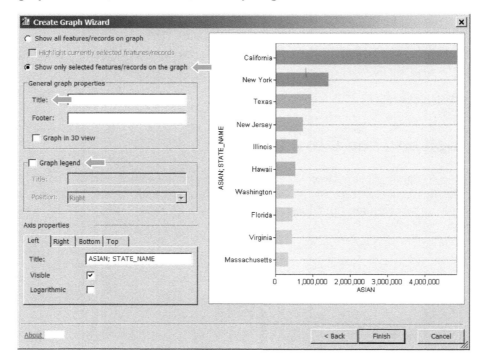

**2**  Click Next and make selections as shown, with "Show only selected features/layers on the graph" marked, Title filled in, and Graph legend unchecked.

**3**   Click Finish.

**4**   Right-click the graph, click Save, browse to Chapter3 of the MyExercises folder, type AsianTop10Graph, and click Save.

**5**   Right-click the graph, click Add to Layout, and close the graph window.

**6**   Snap the lower-left corner of the graph to the intersection of the 0.5-inch vertical and 0.5-inch horizontal guides. Shown is the finished layout, with graph included, in Layout mode.

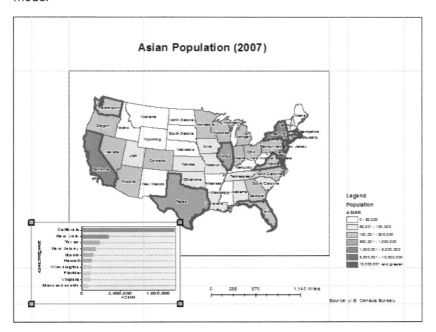

**7**   Save your map document.

## TUTORIAL 3-7
# Building a map animation

Police want to identify new, persisting, and fading spatial clusters of crime locations. These clusters make good areas for police to patrol for enforcement and prevention. Animations of crime data for this purpose are effective if they display two layers of points: new crime points along with recent, but older crime points for context. Then the observer can detect the emergence of new clusters (where there were none before), the persistence of existing clusters getting new crime points, and fading clusters as no new points are added and the cluster fades away. So the approach you'll take calls for two animation layers:one for each day's events and a second with the past two weeks' events. Computer-aided dispatch (CAD) drug calls and shots-fired calls are important crime events for this kind of analysis.

## Open the map document for animation

1   **Open Tutorial3-7.mxd from the Maps folder.** The map document opens to the Middle Hill neighborhood. The CAD data, with dates ranging from 7/1/2012 through 8/31/2012, represents calls from citizens reporting illegal drug dealing and shots fired. The animation that you will build will show the daily sequence of CAD call locations, starting with 7/1/2012.

2   **Save the map document as** Tutorial3-7 **to the Chapter3 folder.**

## Set time properties of a layer

1   **Right-click CAD Calls in the Table of Contents window, click Properties > Time tab, and type or make selections as follows (but do not click OK), with Enable time on this layer checked; Time field set to CALLDATE, Time Step Interval set to 1 day; Time Zone set to Eastern time; and Values are adjusted for daylight savings is checked.** CALLDATE has dates such as 7/1/2012. The Time Step Interval is the unit of time for measurement—here, 1 day.

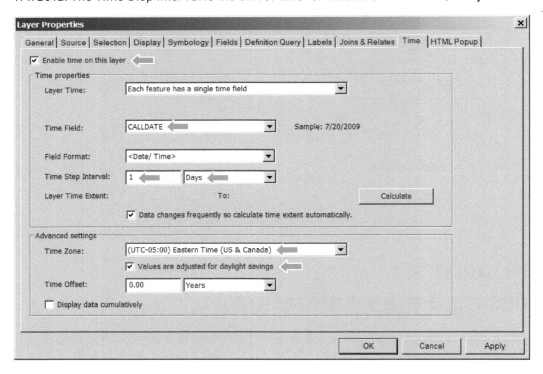

2   **Click Calculate, change the Time Step Interval from 4 to 1, and click OK.**

# Use the Time Slider window for viewing

With time properties set, you are ready to use the Time Slider interface to play a simple video of daily crime points.

**1**   On the Tools toolbar, click the Open Time Slider window button ⬚.

**2**   In the Slider window, click the Options button ⬚ > Playback tab.

**3**   Drag the speed selector to roughly halfway between slower and faster and click OK.

**4**   If needed, click Enable time on the map button in the upper-left corner of the Time Slider window.

**5**   Click the Play button ▶ on the slider. The video plays, one day at a time, shown in the following image at July 30, 2012. Play the video a few more times to see if you can spot any patterns. This is hardly possible until you add the crime time context to the video.

**6**   Close the Time Slider window and save your map document.

# Create a new date column for animating a window of crime points

On any day of the animation, you need to show the current day's crimes with point markers in bright colors and the crime context consisting of two weeks' crime points ending on the same day with black point markers. Displaying a moving window of crime points for two weeks' data requires starting and ending dates for the window. For this purpose, you'll create a new date column with 14 days added to CALLDATE to yield the end date.

**1** **Right-click CAD Calls Context (the layer with the black point markers) in the Table of Contents window and click Open Attribute Table.**

**2** **Click the Table Options button, select Add Field, type** EndDate **for Name, select Date for Type, and click OK.**

**3** **Right-click the header for EndDate and click Field Calculator.**

**4** **Double-click CALLDATE in the Fields panel, click the + button, and type a** blank space **and** 14 **to yield** [CALLDATE] + 14 **for EndDate's expression.**

**5** **Click OK.** The value of EndDate for the first record is 8/3/2012, 14 days after the record's CALLDATE value of 7/20/2012.

**6** **Close the attribute table.**

## Set advanced time properties of a layer

**1** **From the Table of Contents window, right-click CAD Calls Context, click Properties > Time tab, and type or make selections as shown, with Enable time on this layer checked; Layer Time set to Each layer has a start and end time field; Start Time Field set to CALLDATE; End Time Field set to EndDate; Time Step Interval set to 1 day; and Time Zone set to Eastern time.**

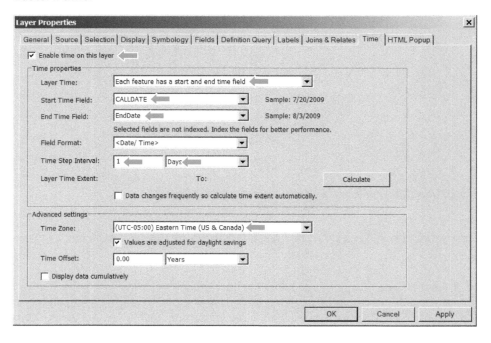

**2** **Click Calculate, change the Time Step Interval from 4 to 1, and click OK.** When you play the animation next, both layers with time properties will animate. CAD Calls will show the current day's crime locations while CAD Calls Context will show all crimes in the interval, including two weeks ending on the current day.

## Use the Time Slider window for advanced viewing

**1** On the Tools toolbar, click the Open Time Slider Window button.

**2** **Turn on the CAD Calls Context layer. Click the Play button on the slider.** Shown is the animation at July 30, 2012. You see all the crime locations for the last two weeks with the last day's crimes in red or yellow and older crimes in black. Currently, there is much persistence in crime clusters.

**3** Save your map document.

## TUTORIAL 3-8
# Using ArcGIS Online

ArcGIS Online (www.arcgis.com) can be used for obtaining, storing, sharing, and using map layers in a browser. Esri sells subscription accounts for ArcGIS Online to organizations but also provides free, public accounts to individuals including 2 GB of free storage for online maps. Thus you can prepare spatial data using ArcMap and ArcCatalog, upload your spatial data to ArcGIS Online, integrate your data with Esri- and user-supplied map layers available from ArcGIS Online, and then share results over the Internet using ArcGIS Online as your free web service host.

## Learn about ArcGIS Online

To learn how to use ArcGIS Online, go to the quick lesson in ArcGIS Learn, "Get started with ArcGIS Online," https://learn.arcgis.com/en/projects/get-started-with-arcgis-online/.

## ASSIGNMENT 3-1
# Create a dynamic map of historic buildings in downtown Pittsburgh

Walking tours are great attractions for tourists, as are historic sites and buildings. Here you'll create a dynamic map document—using visible scale ranges, hyperlinks, and map tips—for the Central Business District of Pittsburgh and some of its historic buildings.

The ArcMap document that you build can be considered a prototype for an ArcGIS Online map, available on the Internet.

While not a part of this assignment, you could export all map layers to shapefile format, compress each shapefile, and upload them to ArcGIS Online to rebuild the map there.

## Get set up

- Rename the folder \EsriPress\GIST1\MyAssignments\Chapter3\Assignment3-1YourName\ to your name or student ID. Store all files that you produce for this assignment in this folder.
- Create a new map document named **Assignment3-1YourName.mxd** with relative paths.

# Build the map

Add the following to your map document:

- \EsriPress\GIST1\Data\Pittsburgh\CBD.gdb\Outline—polygon feature of Pittsburgh's Central Business District neighborhood outline.
- \EsriPress\GIST1\Data\Pittsburgh\CBD.gdb\Bldgs—polygon features of Pittsburgh's Central Business District buildings. Attributes include: Name = name of the building (historic buildings only); Historic = 1 if an historic building, 0 otherwise; Hyperlinks = hyperlinks to websites for historic buildings only; and Address (historic buildings only).
- \EsriPress\GIST1\Data\Pittsburgh\CBD.gdb\Streets2—line features of CBD streets.
- \EsriPress\GIST1\Data\Pittsburgh\CBD.gdb\Curbs—line features of CBD curbs.
- \EsriPress\GIST1\Data\Pittsburgh\CBD.gdb\Histsites—polygon features of historic areas in Pittsburgh's Central Business District.

# Requirements

- Use good labels with halos for streets, historic districts, and historic buildings.
- Symbolize curbs with a medium to dark gray and display them at all scales.
- Give streets no color (so that they are invisible) but label them. Make the labels visible when zoomed in to the largest historic area.
- Label historic areas at all scales.
- Include two copies of buildings in the Table of Contents window. Use a definition query to make one copy historic buildings and the other nonhistoric buildings. Symbolize buildings so that nonhistoric buildings are in ground and historic buildings are figure (use a brown color fill).
- Display the historic buildings at all scales and nonhistoric buildings only when zoomed in to the largest historic area. Label historic buildings with Name when zoomed in to the largest historic district level.
- Use hyperlinks for buildings. *Hint:* Use the Display tab of the property sheet and the Hyperlinks panel there to turn on hyperlinks.
- Use Address for historic building map tips.
- Turn off all useless fields for all map layers. Add aliases if necessary.

# Create a layout comparing 2010 elderly and youth population compositions in Orange County, California

In this assignment, you create a map layout with two maps, one with population fraction of the elderly and the other with population fraction of youths in Orange County, California. The pattern of interest to show is that the fraction of youth population has a higher mean and variance than the fraction of elderly population.

## Get set up

- Rename the folder \EsriPress\GIST1\MyAssignments\Chapter3\Assignment3-2YourName\ to your name or student ID. Store all files that you produce for this assignment in this folder.
- Create a new map document named **Assignment3-2YourName.mxd** with relative paths.

## Create a map layout

Add the following to your map document:

- \EsriPress\GIST1\Data\UnitedStates.gdb\CAOrangeCountyTracts—polygon features for Orange County, California census tracts, Census 2010 with selected census variables: PopTot = total population, Pop0To17 = population ages 17 and younger, and Pop65Up = population 65 and older.

## Requirements

- Include an 8.5-by-11-inch portrait layout with two data frames: one with the fraction of 2010 population 0–17 years old and the second with the fraction 65 or older. **Hint:** Use the Symbology tab in Layer Properties to show a population as a fraction of the total population. Use PopTot field to normalize the data.
- Use the same numeric and color scale for both maps. Use one legend entry for the two choropleth maps.
- Make both data frames the same size. You need to make both data frames in the layout identical in size and alignment. **Hint:** After approximately aligning and sizing the two frames in the layout, right-click a data frame's name in the Table of Contents window and click Properties > Size and Position. Change the frame's width and height to "nice" numbers, for example from 4.9781 to 5.0, and make note of them. Then change the other data frame's width

and height to the same values. Use a vertical guide line in the layout to align the two frames on the left.
- Zoom in to the same populous northwestern half of the county in both frames. *Hint:* Use the approach in tutorial 3-4 with extent indicators to get both frames to have the same extent.
- Include a title, graphic scale bar in miles, and your name as map author.
- Export the map as a JPEG file named **Assignment3-2YourName.jpg**.

## ASSIGNMENT 3-3
# Create an animation for an auto theft crime time series

Auto thieves are often creatures of habit; they return to the same areas and repeat patterns that led to successful thefts in the past. An animation of successive auto theft locations can thus help determine the space-time pattern of such a thief. Suppose that police suspect a serial auto thief who steals cars for basic transportation and then vandalizes abandoned stolen cars in a unique way with spray paint.

## Get set up

- Rename the folder **\EsriPress\GIST1\MyAssignments\Chapter3\Assignment3-3YourName\** to your name or student ID. Store all files that you produce for this assignment in this folder.
- Create a new map document named **Assignment3-3YourName.mxd** with relative paths.

## Create an animation of serial auto thefts

Add the following to your map document:

- \EsriPress\GIST1\Data\DataFiles\AutoTheftCrimeSeries.shp—point features of the suspected crime series of auto thefts.
- \EsriPress\GIST1\Data\Pittsburgh\MidHill.gdb\Outline—polygon feature of the Middle Hill neighborhood boundary.
- \EsriPress\GIST1\Data\Pittsburgh\MidHill.gdb\Streets—line features of streets in the Middle Hill neighborhood.
- \EsriPress\GIST1\Data\Pittsburgh\MidHill.gdb\Curbs—line features of curbs in the Middle Hill neighborhood.

# Requirements

- Use a thick outline for the study area outline, use no color for Streets (so that they are invisible) but label them with a medium-gray font and use the same medium gray for Curbs. Add two copies of AutoTheftCrimeSeries.shp, one named **Auto Theft Crime Series** with a size 10 Circle 2 point marker and bright color fill, and the second named **Auto Theft Crime Series Context** with the same point marker but black color fill and placed under the first copy in the table of contents.
- Set the time properties for the two copies of AutoTheftCrimeSeries like those in tutorial 3-7. Have Auto Theft Crime Series display a single day's auto thefts in each frame and have Auto Theft Crime Series Context display the cumulative set of crimes. Set the playback speed to a medium level.
- Use the Export to Video button on the Time Slider window to create a movie file, **Assignment3-3YourName.avi**. Try playing the movie by double-clicking it in File Explorer.

# 4

# File geodatabases

**ArcGIS can directly use or import most GIS file formats in common use for geo-processing and display. The recommended native file format for use in ArcGIS Desktop is the file geodatabase. It stores map layers, data tables, and other GIS file types in a system folder that has the suffix extension .gdb in its name. In this chapter, you'll learn about working with file geodatabases.**

### Learning objectives

- Build file geodatabases
- Use ArcCatalog utilities
- Modify attribute tables
- Join tables
- Aggregate data

## TUTORIAL 4-1
# Building a file geodatabase

A file geodatabase is simple and flexible, as it is merely a collection of files in a file folder. Nevertheless, a file geodatabase provides a powerful structure and organization for spatial data. You need a special utility program to build and maintain a file geodatabase, ArcCatalog, which you use next. Some of the functionality of ArcCatalog is also available in the ArcMap Catalog window. The Catalog window allows you to do most data utility work while in ArcMap without opening the separate application program, ArcCatalog.

## Open ArcCatalog

**1**  On the Windows taskbar, click Start > ArcGIS > ArcCatalog 10.8.*x*.

**2**  Click the Connect To Folder button ![icon] on the Standard toolbar, expand the folder and file tree under Computer for EsriPress, click the GIST1 folder icon to select it, and click OK. The Connect To Folder button is important to remember. It sets roots for the folder and file tree that you use in ArcCatalog, providing access to files.

## Create a new file geodatabase

While file geodatabases are only folders, you must create them using ArcCatalog or Catalog. Windows Explorer or Computer, while capable of creating a folder, cannot build a file geodatabase.

**1**  In the Catalog Tree panel, expand Folder Connections, EsriPress > GIST1, and the MyExercises folder.

**2**  Click the Chapter4 folder to display its contents (now empty) in the right panel of ArcCatalog.

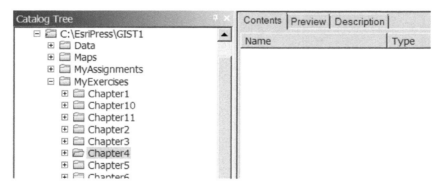

**3**  Right-click Chapter4 in the left panel and click New > File Geodatabase.

**4**  Change the name from New File Geodatabase.gdb to Chapter4.gdb. ArcCatalog creates a file geodatabase that you can populate with feature classes, stand-alone tables, and other objects. Feature classes are map layers stored in a geodatabase.

Next, you import map layers in shapefile format into your new file geodatabase as feature classes.

## Import shapefiles

A shapefile is an older Esri file format that many spatial data suppliers still use for distributing map layers. ArcCatalog allows you to import shapefiles and other map layer formats into a file geodatabase.

**1** In the ArcCatalog right panel, right-click Chapter4.gdb, click Import > Feature Class (multiple). The multiple import option provides the convenience of importing several features at the same time as well as automatically naming output feature classes with the shapefiles' names.

**2** In the Feature Class to Geodatabase (multiple) dialog box, click the browse button to the right of the Input Features field, browse to Data > MaricopaCounty, double-click to open that folder, hold the SHIFT key down, and select tl_2010_04013_cousub10.shp and tl_2010_04013_tract10.shp.

**3** Click Add. That action adds tl_2010_04013_cousub10.shp and tl_2010_04013_tract10.shp to the input panel.

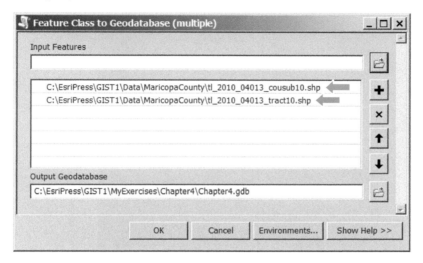

**4** Click OK. ArcCatalog imports the shapefiles into the file geodatabase.

**5** In the Catalog tree, expand the Chapter4 folder and Chapter4.gdb to see the imported feature classes.

## Import a data table

If you have difficulty importing the CensusData$ sheet of CensusData.xlsx (which is the case with some computers), then you'll have to open CensusData.xlsx in Microsoft Excel, export the CensusData$ sheet as text, and add the text file as input rows instead of the spreadsheet.

**1** Right-click the Chapter4 file geodatabase, then click Import > Table (single).

**2** In the Table to Table dialog box, click the browse button to the right of Input Rows, browse to the Data > MaricopaCounty folder, click CensusData.xlsx > Add > CensusData$ > Add. If you have difficulty importing the Excel table, instead copy, save, and import a comma-separated-value file, CensusData.csv, from the same folder.

**3**  Type CensusData **in the Output Table field.** The Field Map panel listing the data fields that are imported may vary.

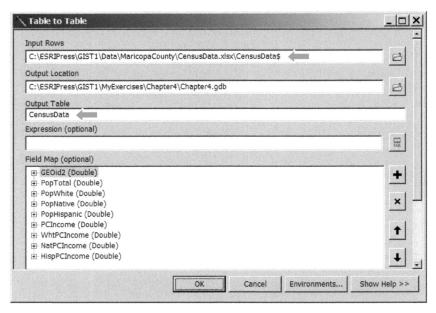

**4**  Click OK.

## TUTORIAL 4-2
# Using ArcCatalog utilities

Now that you've created a file geodatabase, you can start using ArcCatalog utilities. First is Preview, which gives you a good overview of a feature layer or table.

## Preview layers

**1**  If necessary, click Chapter4.gdb in the Catalog tree to expose its contents in the right panel.

**2**   In the right panel, click tl_2010_04013_cousub10 and click the Preview tab. ArcCatalog previews the map layer's geography.

**3**   At the bottom of the Preview window, select Table as the Preview. ArcCatalog previews the map layer's attribute table.

| OBJECTID * | Shape * | STATEFP10 | GEOID10 | NAME10 |
|---|---|---|---|---|
| 1 | Polygon | 04 | 0401390459 | Buckeye |
| 2 | Polygon | 04 | 0401390561 | Chandler |
| 3 | Polygon | 04 | 0401390867 | Deer Valley |
| 4 | Polygon | 04 | 0401391377 | Gila Bend |
| 5 | Polygon | 04 | 0401392601 | Phoenix |
| 6 | Polygon | 04 | 0401393009 | St. Johns |
| 7 | Polygon | 04 | 0401393060 | Salt River |
| 8 | Polygon | 04 | 0401393472 | Tonto National Forest |
| 9 | Polygon | 04 | 0401393774 | Wickenburg |

**4**   Click the Description tab. ArcCatalog previews the map layer's metadata in a report format. For TIGER maps, only general metadata is available.

> **Description**
> The TIGER/Line Files are shapefiles and related database files (.dbf) that are an extract of selected geographic and cartographic information from the U.S. Census Bureau's Master Address File / Topologically Integrated Geographic Encoding and Referencing (MAF/TIGER) Database (MTDB). The MTDB represents a seamless national file with no overlaps or gaps between parts, however, each TIGER/Line File is designed to stand alone as an independent data set, or they can be combined to cover the entire nation.

**5**   Click the Contents tab.

## YOUR TURN

Preview tl_2010_04013_tract10 and CensusData.

# Rename feature layers

Because a file geodatabase has a special file format, you must use ArcCatalog for file-management purposes, including renaming and copying items.

**1** **In the left panel under Chapter4.gdb, right-click tl_2010_04013_cousub10, click Rename, type** Cities, **and press the Tab key.**

**2** **Repeat step 1 to rename CensusData to** CensusTractData, **and tl_2010_04013_tract10 to** Tracts.

# Copy and delete feature layers

**1** **In the left panel under Chapter4.gdb, right-click Cities, click Copy, right-click Chapter4.gdb, click Paste > OK.** ArcCatalog creates the copy, Cities_1.

**2** **Right-click Cities_1 and click Delete > Yes.**

> ## YOUR TURN
> Open File Explorer, browse to Chapter4.gdb in the Chapter4 folder of MyExercises, right-click Chapter4.gdb to get its properties and size, and take a look at the files inside of it comprising the cities, tracts, and census tract data. Notice that the folder size is approximately 1.52 MB on the disk and that its approximately 53 files are incomprehensible. You need ArcCatalog or Catalog in ArcMap to use, manipulate, and understand these files. Leave File Explorer open for use in the following steps.

# Compress a file geodatabase

**1** **In the left panel of ArcCatalog, right-click Chapter4.gdb, click Administration > Compress File Geodatabase > OK.**

**2** **Use File Explorer to check the size of the Chapter4.gdb folder.** Verify that the compressed file geodatabase is about 1 MB on your disk, around 50 files, and about a 40 percent reduction in size.

Although ArcMap can process compressed feature layers by uncompressing them on the fly, you will use the next step to uncompress the folder and get the layers back to original size.

**3** **In the left panel, right-click Chapter4.gdb, click Administration > Uncompress File Geodatabase > OK.**

**4** **Close ArcCatalog.**

## YOUR TURN

Open ArcMap and create a new map document named Tutorial4-3.mxd with relative paths stored in the Chapter4 folder of MyExercises. Add the two layers and table from Chapter4. gdb and symbolize as seen in the figure. Save your map document.

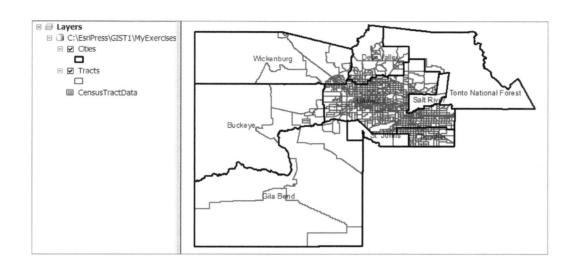

## TUTORIAL 4-3
# Modifying an attribute table

Most of what gets processed or displayed in a GIS depends on attribute table values. There is much, therefore, that you need to know about working on tables, including deleting, modifying, and creating columns.

## Delete unneeded columns

Many map layers have extra or unnecessary attributes, from the user's point of view, that you can delete to tidy up.

1   **In the table of contents, right-click Tracts and click Open Attribute Table.** It's not possible to delete the primary key created by ArcGIS, ObjectID, or the Shape attribute. GEOID10 is the tract geocode that you need for joining the CensusTractData table to the tracts map layer, but all other fields are candidates for deletion.

**2**  In the table, right-click the header for the STATEFP10 column, click Delete Field, and click Yes.

**3**  Similarly delete the following fields: TRACTCE10, NAME10, NAMELSAD10, MTFCC10, and FUNCSTAT10.

**4**  Close the Tracts table.

---

## YOUR TURN

GEOID10 is a unique city identifier in the United States. Delete fields from the Cities layer to keep just OBJECTID, Shape, GEOID10, NAME10, Shape_Length, and Shape_Area. When finished, close the Cities table.

---

## Modify a geocode

It is often necessary to join two tables to make a single combined table. For example, there are thousands of census variables, so it is impractical to have all needed census variables for tracts stored in the Tracts attribute table. Instead, you'll select the variables you want, download a corresponding table (which includes tract geocodes) from the Census Bureau website, and join the table to the tract polygon table by geocode. The GEOID10 column of the Tracts attribute table and the GEOid2 column of the CensusTractData table are the corresponding geocodes for these tables. These attributes would match, except that GEOID10 has a text data type whereas GEOid2 has a numeric data type. You can tell because text data is left-aligned in its column while numeric data is right-aligned.

**1**  In the table of contents, right-click CensusTractData, click Open, and sort GEOid2 ascending. You can see that this numeric attribute's first sorted value is 4013010101.

**2**  Close CensusTractData. Similarly, open the Tracts attribute table and sort GEOID10 ascending. The first sorted value is the text value 04013010101, with a leading zero, but otherwise matches the geocode in CensusTractData.

To get this table's geocode to join with that in CensusTractData, all you have to do is create a numeric version of it, which is easy with the Field Calculator. The numeric version drops the leading zero.

**3**  In the Tracts table, click the Table Options arrow and click Add Field.

**4**  In the Fields dialog box, type GEOID10Num in the Name field, change the Type to Double, and click OK. You need Double, which can have values up to 15 digits in length, to store GEOID10.

**5**  Right-click the GEOID10Num header and click Field Calculator > Yes.

**6** In the Field panel, double-click GEOID10 (to set GEOID10Num = GEOID10), and click OK. That provides the needed geocode for joining the CensusTractData table to the Tracts attribute table.

**7** Close the Tracts table and open the CensusTractData table.

Before making the join, first you will create some new calculated columns in the CensusTractData table needed for a map document.

## Calculate a new column

The CensusData table has population and per capita income for the total population, whites, Native Americans, and Hispanics. Desired for mapping are two ratios of per capita incomes for Native Americans divided by whites and for Hispanics divided by whites.

**1** In the CensusTractData table, click the Table Options arrow and click Add Field.

**2** In the Add Field window, type RNatWht in the Name field, change the Type to Float, and click OK.

The new column will contain the ratio of Native American per capita income to white per capita income. Wherever this ratio is greater than one, Native Americans on average earn more than whites. First, however, you must select only records where PCIncWht is greater than zero, because PCIncWht is the divisor for this ratio and is used to calculate values for RNatWht. Anything divided by zero is undefined, so this case must be avoided; this is done through the selection.

**3** In the CensusTractData table, click the Table Options arrow and click Select By Attributes.

**4** On the Select By Attributes dialog box, scroll down the list of fields, double-click WhtPCIncome to add it to the lower Select panel, click the > Symbol button, click Get Unique Values, and double-click 0 in the Unique Values list. Those actions create the expression WhtPCIncome > 0.

**5** Click Apply.

**6** Right-click the RNatWht header and click Field Calculator > Yes.

**7**  In the Field Calculator window, double-click NatPCIncome in the Fields panel, click the / button, double-click WhtPCIncome in the Fields panel, and click OK. Rows with values for both the numerator and denominator get data values, those with a denominator that is positive but no numerator get a zero, and those with 0 or null for the denominator retain the null value.

## YOUR TURN
Repeat the previous steps to calculate a new column in the CensusTractData table named RHisWht, which is the ratio of HispPCIncome divided by WhtPCIncome. This is the ratio of per capita income of Hispanics divided by the per capita income of whites. Clear the selection and close the table when finished. Save your map document.

## TUTORIAL 4-4
# Joining tables

Next, you will join the CensusTractData table to the polygon Tracts feature class. The same steps work if your map layer is a shapefile or map layer in another format supported by ArcMap.

**1**  Save your map document as Tutorial4-4.mxd to the Chapter4 folder.

**2**  In the ArcMap Table of Contents, right-click the Tracts layer, click Joins and Relates > Join.

**3**  Make the selections shown in the figure, with Join attributes from a table in the top box, GEOID10Num for no. 1, CensusTractData for no. 2, and Geoid2 for no. 3. Click OK.

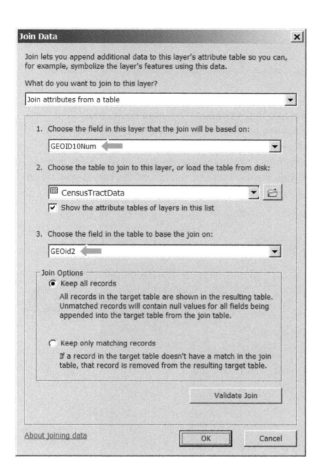

**4**  Open the Tracts attribute table, scroll to the right in the table, and verify that ArcMap joined the CensusTractData table to the Attributes of Tracts table.

| | Shape_Length | Shape_Are | GEOID10Num | OBJECTID * | GEOid2 * | PopTotal | PopWhite | PopNative |
|---|---|---|---|---|---|---|---|---|
| | 0.127403 | 0.000755 | 4013422644 | 712 | 4013422644 | 7287 | 6319 | 33 |
| | 0.09817 | 0.000499 | 4013422643 | 711 | 4013422643 | 5789 | 4929 | 41 |
| | 0.063555 | 0.000249 | 4013422642 | 710 | 4013422642 | 5715 | 4888 | 41 |
| | 0.132432 | 0.000748 | 4013422641 | 709 | 4013422641 | 6346 | 5052 | 54 |
| | 0.170811 | 0.001141 | 4013815900 | 889 | 4013815900 | 5981 | 4533 | 23 |
| | 0.075694 | 0.000274 | 4013815800 | 888 | 4013815800 | 4083 | 3279 | 45 |

Tracts

**5** Close the Tracts table.

---

## YOUR TURN

Now that you have census tract data attached to the census tract map layer, you can make a quick map of an interesting variable. Are there places in Maricopa County where Native Americans typically have higher per capita income than whites? Make a choropleth map using the RNatWht (ratio of Native American per capita income divided by white per capita income) attribute you created. Use quantiles for the numeric scale and the red-to-yellow-to-green color ramp. When finished, save your map document.

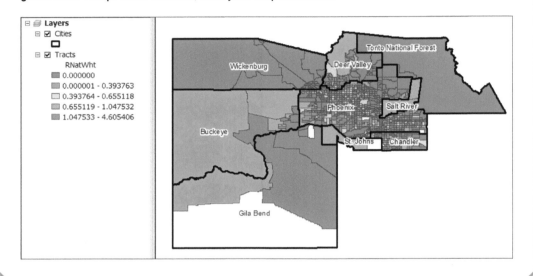

---

## TUTORIAL 4-5
# Creating centroid coordinates in a table

The centroid of a polygon is the point at which the polygon would balance on a pencil point if it were cut out of cardboard. A map layer consisting of polygon centroid points has many uses; for example, together, polygons and their centroids give you the ability to display two attributes of the same map layer in the same map, one as a choropleth map and the other as a size-graduated point marker map. Next, you'll use the ArcMap tool for calculating polygon centroids.

# Add x,y coordinates to a polygon attribute table

As a precaution and as a simplification, when you add new fields to an attribute table that has a joined table, it's a good idea to remove the join first, create the new fields, and then rejoin the table.

**1** **Save your map document as** Tutorial4-5.mxd **to the Chapter4 folder.**

**2** **Right-click Tracts in the table of contents and click Joins and Relates > Remove Joins > Remove All Joins.** Your choropleth map disappears because the attribute that was displayed was in the joined table, which you just removed.

**3** **Open the Tracts table, click the Table Options arrow, and click Add Field.**

**4** **Type** X **as the Name, select Double for Type, and click OK.**

**5** **Repeat steps 3 and 4 except call the new field** Y.

**6** **Right-click the X header and click Calculate Geometry > Yes. Examine the Calculate Geometry window and click OK > Yes.**

**7** **Repeat step 6 except right-click the Y header and select Y Coordinate of Centroid. Close the attribute table when finished.**

| Shape_Length | Shape_Area | GEOID10Num | X | Y |
|---|---|---|---|---|
| 0.127403 | 0.000755 | 4013422644 | -111.598551 | 33.366828 |
| 0.09817 | 0.000499 | 4013422643 | -111.618524 | 33.357267 |
| 0.063555 | 0.000249 | 4013422642 | -111.627432 | 33.371808 |
| 0.132432 | 0.000748 | 4013422641 | -111.661659 | 33.357603 |
| 0.170811 | 0.001141 | 4013815900 | -111.757895 | 33.264 |
| 0.075694 | 0.000274 | 4013815800 | -111.697149 | 33.284485 |

## YOUR TURN
Rejoin the CensusTractData table to the Tracts layer.

# Export a table

When you export joined tables as a table, you get all the attributes of the joined tables stored as one table permanently. (Similarly, if you export a map layer that has a joined table, the new map layer's attribute table has both tables joined permanently.) There are several possible uses for the new table, one of which is to use it to make a new point layer based on the centroid coordinates.

**1** **Open the Tracts attribute table and click the Table Options arrow > Export.**

**2**   In the Output table field of the Export Data window, click the browse button, change the Save As type to File Geodatabase tables, browse to the Chapter4 folder, double-click Chapter4.gdb, change Name to TractCentroids, and click Save > OK > Yes. Open the table to see that it has all the columns of both joined tables. Then close the table.

**3**   Close the Tracts attribute table.

## Create a feature class from an XY table

**1**   On the Menu bar, click Windows > Catalog. This opens a version of ArcCatalog as a window in ArcMap, thus providing quick access to GIS utility programs.

**2**   Expand the Chapter4.gdb file geodatabase.

**3**   Right-click TractCentroids, click Create Feature Class > From XY Table.

**4**   In the Create Feature Class From XY Table window, click the Coordinate System of Input Coordinates button, click the Add Coordinate System button 🌐 ▼ > Import, browse to Chapter4.gdb, double-click Tracts in Chapter4.gdb, and click OK.

The coordinate system's geographic spherical coordinates are the same as those of the Tracts layer, so the simplest option is to import the system specification from Tracts.

**5**   Click the browse button for Output, change the Save As type to File and Personal Geodatabase feature classes, browse to and double-click Chapter4.gdb, change Name to CensusTractCentroids, then click Save > OK.

**6**   Drag CensusTractCentroids from the Catalog window to the map document and drop it at the top of the table of contents.

**7**   Close the Catalog window in ArcMap.

**8**   Open the Cities attribute table, click the row selector for Phoenix to select that record and polygon on the map, and close the table.

**9**   Right-click Cities in the table of contents and click Selection > Zoom to Selected Features. Now you can get a better look at the centroids point layer you just created.

## Symbolize a choropleth and centroid map

Here you symbolize a map using both the tract polygons and centroids. Let's see how per capita income compares with percentage of total population that is Hispanic.

**1**  In the table of contents, right-click the Tracts layer, click Properties > the Symbology tab.

**2**  In the Show panel, click Quantities > Graduated Colors. Under Fields, select PopHispanic for the Value field, change Normalization to PopTotal, click Classify, change Method to Quantile, and click OK. Change the color ramp to yellow through brown.

**3**  In the Symbology tab, click the Label header to the right of the Symbol and Range headings, click Format Labels, click the Numeric Category, change the number of decimal places to 2, and click OK twice.

**4**  Right-click CensusTractCentroids in the table of contents, click Properties > the Symbology tab.

**5**  In the Show panel, click Quantities > Graduated Symbols. Under Fields, change Value to PCIncome.

**6**  Change the Symbolize Size range to 2 to 10.

**7**  Click the Template button, choose Circle 2, and click OK.

**8**  In the Classification panel, change the number of classes to 5, click Classify, change Method to Quantile, and click OK.

**9** In the Symbology tab, click the Label header, click Format Labels, click the Numeric Category, change the number of decimal places to 0, click OK twice, and save your map document. Now you can plainly see that Phoenix, Arizona, has areas with concentrations of Hispanic population, and those areas tend to be low income.

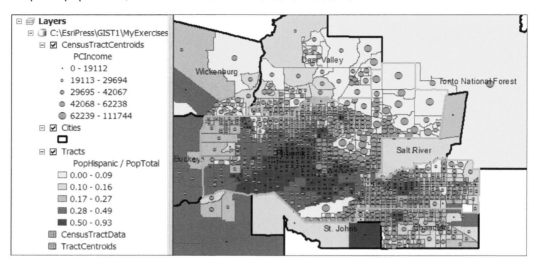

## TUTORIAL 4-6
# Aggregating data

A car beat is the patrol area of a single police car. This tutorial has you count (or aggregate) eating and drinking businesses in Rochester, New York, within car beats. Such businesses are crime attractors, so it is useful to have a map showing where they are concentrated. In this case, the workflow has four steps. First, you'll join a code table to all business points in order to identify the businesses of interest. Second, you apply a definition query to businesses, limiting them to eating and drinking businesses. Third, you use a spatial join of car beats to eating and drinking businesses to count them up by car beat. Finally, you use the results to create a choropleth map of car beats representing the number of drinking and eating places in each car beat.

## Join a code table to an attribute table

Here you do a one-to-many join. Earlier in this chapter, you did a one-to-one join between the CensusTractData table and the Tracts attribute table. Each CensusTractData record had one and only one matching Tract polygon and record. This time, each SIC (standard industrial code) record in the code table potentially has many matching records in the Businesses point layer; for example, all of the drinking places in Rochester for the SIC record for code 5813.

**1**   **Open Tutorial4-6.mxd from the Maps folder.** The map that opens displays police car beats in Rochester as polygons, and all businesses as points.

**2**   **Save your map document to the Chapter4 folder.**

**3**   **Open the Businesses attribute table and examine its attributes.** There are 9,325 businesses of all kinds with the name and address of each business along with a four-digit SIC code. To break this code, you need the code table named SIC. Notice that the SIC column in the Businesses attribute table has the text data type.

| OBJECTID * | Shape * | NAME | ADDRESS | SIC |
|---|---|---|---|---|
| 9008 | Point | WATER-WISE INC | 311 EXCHANGE BLVD | 7389 |
| 9006 | Point | WATER STREET GRILL | 179 N WATER ST | 5812 |
| 9007 | Point | WATER STREET MUSIC HALL | 204 N WATER ST | 5813 |
| 9009 | Point | WATERPRO SUPPLIES CORP | 609 BUFFALO RD | 9999 |
| 9010 | Point | WATERWORKS J WELLINGTONS | 315 ALEXANDER ST | 5812 |
| 9011 | Point | WATKINS ADVERTISING | 105 LANARK CRES | 9999 |

**4**   **Close the Businesses attribute table. Click the List By Source button at the top of the table of contents and open the SIC table.** This code table has two-, three-, and four-digit SIC codes and descriptions. The two-digit code, 01 for Agriculture, is broken down into three-digit categories, for example 011 Cash grains. Then three-digit codes are broken down into four-digit business types, for example, 0111 Wheat. Note that SICCODE also has the text data type, which makes the join to Businesses possible.

| OBJECTID * | SICCODE * | SICDESCR |
|---|---|---|
| 1084 | 5736 | Musical instrument stores |
| 1085 | 58 | Eating And Drinking Places |
| 1086 | 581 | Eating and Drinking Places |
| 1087 | 5812 | Eating places |
| 1088 | 5813 | Drinking places |
| 1089 | 59 | Miscellaneous Retail |

**5**   **Close the SIC table. Right-click the Businesses layer in the table of contents and click Joins and Relates > Join, and type or make the selections shown in the figure, with Join attributes from a table in the top box, SIC for no. 1, SIC for no. 2, and SICCODE for no. 3.**

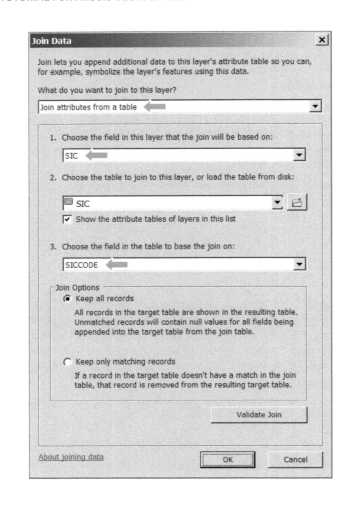

**6**  **Click the Validate Join button.** You get a report that 9,253 of 9,325 business records successfully join. If you needed to diagnose and repair the non-matches, you could, but you skip that here.

**7**  **In the Join Validation window click Close and click OK.**

**8**  **Open the Businesses attribute table, scroll to the right, and see the joined SIC code descriptions.** Of course, all irrelevant SICCODE values, including those two- and three-digit values, were ignored in the join process. Only relevant four-digit codes were used.

**9**  **Close the attribute table.**

## Create a definition query

**1**  **Open the Businesses layer's properties and click the Definition Query tab.**

**2**   **Click the Query Builder button, and build the query criterion:** SIC.SICDESCR = 'Eating places' OR SIC.SICDESCR = 'Drinking places' **using the Get Unique Values button to get description values.**

**3**   **Click OK > OK.** There are 457 such places now displayed.

## Use a spatial join to count eating and drinking places by car beat

If you do a spatial join of a polygon layer with a point layer, the result is statistics by polygon for the points. Each such join automatically includes a count of points per polygon, and if there are quantitative attributes of the points they can have statistics included, as shown in the figure for step 1.

**1**   **Right-click Car Beats, click Joins and Relates > Join, and make selections as shown, with Join data from another layer based on spatial location for the top box, Businesses for no. 1, and** *CarBeatsEatingDrinkingBusinesses* **for no. 3.**

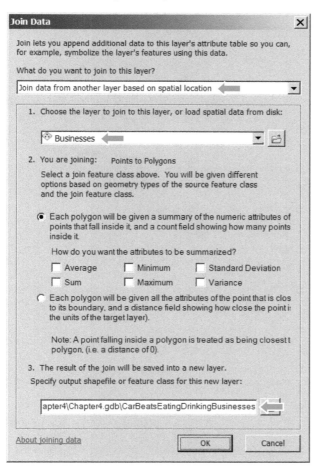

**2**  Click OK.

**3**  Remove the original car beats layer from the table of contents.

**4**  Symbolize the choropleth map using the Count_ attribute of
CarBeatsEatingDrinkingBusinesses. Use five quantiles and label each polygon with the
Beat attribute. The result is a good, high-level map for scanning Rochester for car beats with
many eating and drinking places. A good interactive version of this map would have the new
layer turn off when zoomed in and the Businesses point layer turned on for details.

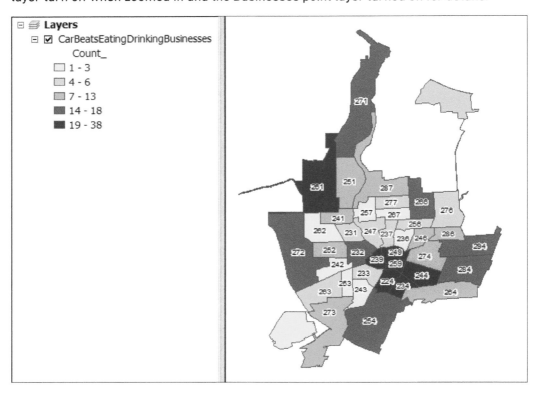

**5**  Save your map document and close ArcMap.

# Investigate educational attainment

In this assignment, you carry out all the standard steps to work with census data in a GIS. Chapter 5 covers downloading and preparing data from the Census website, so this assignment starts at the point after census data is downloaded and ready for use. You create a file geodatabase, import a census tract map and data table into the file geodatabase, carry out several data utility and calculation steps, join the data table to the tract map, and produce a map layout. The data is on education attainment in Allegheny County, Pennsylvania. Your objective is to create a layout comparing the fraction of males versus females 25 or older who have an associate college degree or higher. You can imagine that spatial patterns are similar, but is high educational attainment the same for both sexes?

## Get set up

First, rename your assignment folder and create a map document.

- Rename the folder **\EsriPress\GIST1\MyAssignments\Chapter4\Assignment4-1YourName\** to your name or student ID. Store all files that you produce for this assignment in this folder.
- Create a new map document named **Assignment4-1YourName.mxd** with relative paths.

Use the following steps to project the Table of Contents layer to the state plane coordinate system, with coordinates in feet:

- Right-click Layers in the table of contents and click Properties > Coordinate System tab.
- Expand Projected Coordinate Systems to State Plane > NAD 1983 (US feet) and click NAD 1983 State Plane Pennsylvania South FIPS 3702 (US Feet) > OK.

## Build the map

Create a new file geodatabase named **Assignment4-1YourName.gdb**. Import data into it:

- Import \EsriPress\GIST1\Data\DataFiles\tl_2010_42003_tract10.shp as a feature class and rename it Tracts. GEOID10 is the tract geocode in this table and it has the text data type.
- Import \EsriPress\GIST1\Data\DataFiles\AllCoEdAttain.xlsx, HighEdAttainment$ worksheet as a table and rename it **EducationalAttainment**. This is 2010 census tract data for Allegheny County on educational attainment. Attributes include: GEOid2 = tract geocode, numeric data type; Female25_ = total female population 25 and older; FemaleAssociateDegree = female population 25 or older with an associate degree as highest educational attainment; FemaleBachelorDegree = female population 25 or older with a bachelor's degree as highest educational attainment; and so forth up through FemaleDoctorateDegree. Include FemaleProfessionalDegree in your calculations. Males have similar attributes.

Add Tracts and EducationalAttainment to your map document.

Prepare Tracts for use:

- Create a new geocode, matching the data type of the geocode in the data table.
- Clean up the attribute table, deleting unneeded attributes.

Create two new attributes in EducationalAttainment:

- **PCollegeF** = Percentage of females 25 and older who have an associate college degree or higher (so values possibly could range between 0 and 100). Use Float as the data type.
- **PCollegeM** = Percentage of males 25 and older who have an associate college degree or higher. Use Float as the data type.
- *Hint:* Before calculating PCollegeF, select rows where Female25_ > 0. Otherwise you will get a warning message that the calculation failed (in cases where the denominator of your expression was zero). Do the same for males.

Create a layout that compares the two new attributes, PCollegeF and PCollegeM.

- Use the map layer and table in your file geodatabase (and not the shapefile or Excel table).
- Use two data frames and two separate choropleth maps, one for each new attribute.
- Use the same color ramp and numeric scale for both maps and include a single, common legend.
- *Hint:* Finish one data frame. Then copy and paste it for the second data frame, making modifications as necessary. Create a layer file for the first data frame's map and use it to symbolize the second data frame's map.

Create a scatter plot of PCollegeF versus PCollegeM and add it to the layout.

- Use View > Graphs > Create Graph to create the scatter plot. Experiment with modifying and improving your graph's axes (e.g., use tic marks of 20), add vertical grid lines, increase the sizes of fonts, etc., by right-clicking your graph and clicking Advanced Properties.
- Once the plot is created, resize its window so that the vertical and horizontal axes are the same length.
- Right-click the graph and click Copy As Graphic. On your layout, click Edit > Paste. Reposition and align the graph and other elements of your layout.
- On the layout window, click Customize > Toolbars > Draw and draw a red line from (0,0) to (100,100) on your graph. This is the line of perfect equality of educational attainment for males and females. Think about the pattern you see on your maps and the graph.
- Save your layout and export it as **Assignment4-1YourName.jpg**.

## ASSIGNMENT 4-2

# Compare serious crime with poverty in Pittsburgh

The criminology literature finds that much crime is related to poverty: the larger the population living in poverty, the higher the rate of certain kinds of crimes. Let's see if this relationship is evident in Pittsburgh at the census tract level. In the process, you'll build a file geodatabase, import map layers into it, build a code table from scratch and join it to a map layer, join census data to a map layer, and carry out a spatial join to aggregate data.

## Get set up

First, rename your assignment folder and create a map document.

- Rename the folder **\EsriPress\GIST1\MyAssignments\Chapter4\Assignment4-2YourName\** to your name or student ID. Store all files that you produce for this assignment in this folder.
- Create a new map document named **Assignment4-2YourName.mxd** with relative paths.

Use the following steps to project the Table of Contents layer to the state plane coordinate system, with coordinates in feet:

- Right-click Layers in the table of contents, click Properties > Coordinate System tab.
- Expand Projected Coordinate Systems > State Plane > NAD 1983 (US feet) and click NAD 1983 State Plane Pennsylvania South FIPS 3702 (US Feet) > OK.

## Build the map

Create a new file geodatabase named **Assignment4-2YourName.gdb**. Import the following data into it:

- \EsriPress\GIST1\Data\Pittsburgh\Shapefiles\PittsburghSeriousCrimes2008.shp—point shapefile of serious crime offense locations in Pittsburgh during summer, 2008. Attributes include: CCN = police ID for offense, Address = location of the offense, DateOccur = date of the offense, Hierarchy = FBI hierarchy code for the offense (1 = Criminal Homicide, 2 = Forcible Rape, 3 = Robbery, 4 = Aggravated Assault, 5 = Burglary, 6 = Larceny-theft, 7 = Motor Vehicle Theft, 8 = Arson).
- \EsriPress\GIST1\Data\Pittsburgh\City.gdb\PghTracts—polygon features of Pittsburgh 2010 census tracts.
- \EsriPress\GIST1\Data\Pittsburgh\Shapefiles\PovertyTracts.xlsx with PghPovertyTracts worksheet—2010 tract data for poverty. Attributes include: GEOid = tract geocode (numeric data type), PopWithPovStatus = population for whom poverty status is known (presumably

the same as total population), PopBelowPovLevel = population who are below the poverty level.

Join PghPovertyTracts to PghTracts and make a choropleth map for population below the poverty line.

Create a new code table in your file geodatabase:

- Right-click your file geodatabase in Catalog and click New > Table.
- Name the table **UCRHierarchy** with no alias. Click Next > Next.
- Create a field named **UCR** with the Short Integer data type. Create another field named **Crime** with the Text data type and length 20.
- Click Customize > Toolbars > Editor.
- On the editor toolbar, click Editor > Start Editing > Continue.
- Open UCRHierarchy in the table of contents, type **1** for UCR and **Criminal Homicide** for Crime in the first row.
- Finish up inputting rows found with codes and crime types from the description of PittsburghSeriousCrimes2008.shp above. When done, click Editor in the Editor toolbar, Save Edits, Stop editing, and close the Editor toolbar. Close your new table.

Create the following query definition for PittsburghSeriousCrimes2008 and rename that layer in the table of contents to **PittsburghSeriousCrimesSummer2008**:

"DateOccur" >= date '2008-06-01' AND "DateOccur" <= date '2008-08-31'.

Join UCRHierarchy to PittsburghSeriousCrimesSummer2008. Symbolize PittsburghSeriousCrimesSummer2008 with the Crime attribute using unique values. Use different shapes and colors for point markers. This layer will display only when you zoom in.

Spatially join PittsburghSeriousCrimesSummer2008 to PghTracts to get a count of serious crimes per tract in summer 2008. Name the output **CrimeAggregatedByTracts**. Use the results to symbolize PghTracts centroids with size-graduated point markers. *Hint:* Use Quantities, Graduated symbols for symbolizing the crime count (there is no need to create tract centroids in this case).

Set threshold scales so that when zoomed in to about a fourth of Pittsburgh or farther, PittsburghSeriousCrimesSummer2008 turns on and CrimeAggregatedByTracts turns off.

Create a layout of your own design for the map document, zoomed to full extent.

If you have difficulty carrying out the spatial join of PittsburghSeriousCrimesSummer2008 to PghTracts, do the following.

- Right-click PghTracts and click Data > Export Data.
- Save the exported data and add to your map as CrimeAggregatedByTracts.
- Then spatially join PittsburghSeriouscrimesSummer2008 to CrimeAggregatedByTracts.

# Analyze the data

Get set up to create a graph:

- Open the CrimeAggregatedByTracts property table, click the Fields tab, and give Count_ the alias Serious Crimes.
- Open the PghPovertyTracts table and give PopBelowPovLevel the alias Population Below Poverty Level.
- Join PittsburghPovertyTracts to CrimeAggregatedByTracts.

Create a scatter plot of Serious Crimes versus Population Below Poverty Level:

- Use View > Graphs > Create Graph to create the scatter plot.
- Once created, resize its window so that the vertical and horizontal axes are the same length.
- Right-click the graph and click Copy As Graphic. On your layout, click Edit > Paste. Reposition and align the graph and other elements of your layout.
- Save your layout and export it as **Assignment4-2YourName.jpg**.

# 5

# Spatial data

Vast collections of spatial data are available from government agencies. You can readily download much of this data for free on the Internet, but before doing so it is helpful to get some background and see some of the major forms of this data. Spatial data is complex, with both vector and raster formats available in many file formats and with several attending characteristics such as coordinate system, feature or cell attribute properties, and intended map scale for application. This chapter provides a hands-on introduction to spatial data, and then has you download or use samples from some of the major governmental suppliers.

Learning objectives

- Learn about metadata
- Work with world projections
- Work with US map projections
- Work with systems of projections
- Learn about vector data formats
- Download US Census map layers
- Download US Census data
- Download American Community Survey data
- Explore web services for raster maps
- Use nationalatlas.gov

## TUTORIAL 5-1
# Examining metadata

Spatial data needs much documentation for interpretation and proper use. Metadata is the term for such documentation. It describes the context, content, and structure of GIS data. In ArcGIS Desktop, data providers use ArcCatalog to create and view metadata.

## Open a map document

**1** **Start ArcMap and open Tutorial5-1.mxd from the Maps folder.** These map layers come from two different data sources. The Pennsylvania Counties and Allegheny Tracts layers were obtained from the US Census Bureau, and the Allegheny County Municipalities layer was obtained from a local GIS organization, the Southwestern Pennsylvania Commission.

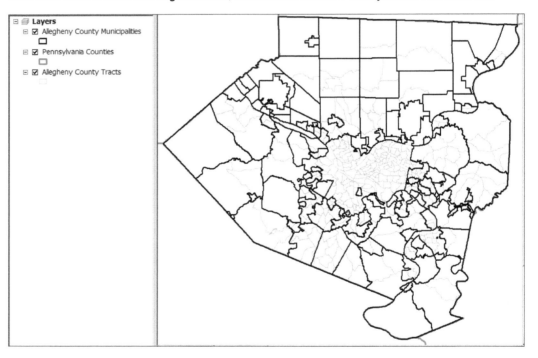

**2** **Save the map document to the Chapter5 folder of MyExercises.**

## Open a metadata file

ArcGIS metadata is described in a metadata file that can be read using ArcCatalog.

**1** **On the taskbar, click Start > ArcGIS > ArcCatalog 10.8.x** Depending on your operating system and how ArcGIS and ArcCatalog have been installed, you may have a different navigation menu.

**2** **Click Customize > ArcCatalog Options and the Metadata tab.**

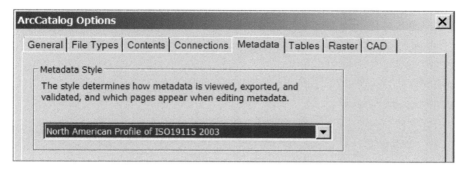

**3** **Click North American Profile of ISO19115 2003 as the Metadata style > OK.**

**4** **In the Catalog tree, browse to UnitedStates.gdb in the Data folder, click PACounties, and the Description tab.** Be patient. It takes time for the description to appear.

**5** **Scroll to the Spatial Reference option and read the entry.** You learn that geographic coordinates are spherical, Geographic Coordinate System, NAD 1983 Datum. Note: Scroll down in the Description tab to see the spatial reference data. If you do not see any metadata, click the Contents tab and then press the Description tab again.

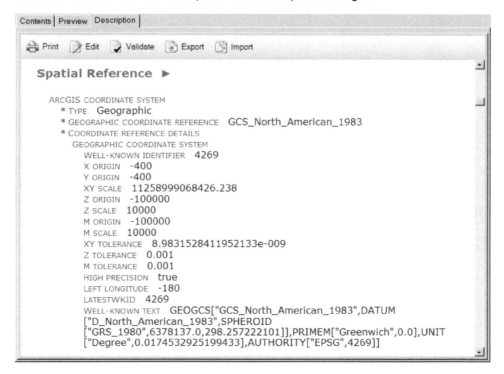

**6**  In the Catalog tree, browse to AlleghenyCounty.gdb in the Data folder, click Munic, click the Description tab, and wait while it loads. If you get an Update Metadata window, click Yes > OK and click Munic in the data tree again.

**7**  Scroll to the Spatial Reference option and read the entry. You learn that the layer's projection is rectangular, NAD_1983_StatePlane_Pennsylvania_South_FIPS_3702_Feet. You will learn more about projections and coordinate systems in tutorials 5-2 through 5-4.

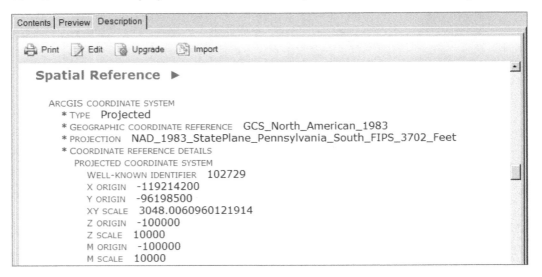

**YOUR TURN**
Explore some of the other metadata for Pennsylvania Counties, Allegheny Municipalities, and Allegheny County Tracts 2010, including source credits, fields, and other spatial information. When finished, close ArcCatalog.

## TUTORIAL 5-2
# Working with world map projections

There are two types of coordinate systems—geographic and projected. Geographic coordinate systems use latitude and longitude coordinates for locations on the surface of a sphere while projected coordinate systems use a mathematical transformation to a flat surface and rectangular coordinates.

# Set world projections

ArcMap has more than 100 projections from which you may choose. Typically, though, only relatively few are needed for most purposes.

**1**  **Open Tutorial5-2.mxd from the Maps folder and save the map document to the Chapter5 folder.**

**2**  **Place your cursor over the westernmost point of Africa and read the coordinates on the bottom of the ArcMap window (approximately –17, 20 decimal degrees).** The map and data frame are in geographic coordinates, decimal degrees, which are angles of rotation of the Earth's radius from the prime meridian at the equator. These coordinates are not intended for viewing on a flat surface such as your computer screen, so there are large distortions if you do so; for example, the north and south pole points are the horizontal lines bounding the top and bottom of the map. Instead of geographic coordinates, you should use one of the projected coordinate systems appropriate for viewing flat maps of the world. ArcMap has several and can easily project the map on the fly.

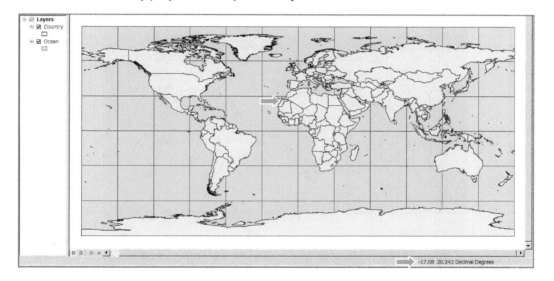

# Change the map's projection to Mercator

**1**  **In the table of contents, right-click Layers, click Properties > Coordinate System tab.**

**2**  **In the Select a coordinate system panel, scroll down to and click the plus (+) sign beside Projected Coordinate Systems, and click the plus sign beside the World folder.**

**3**  **Scroll down the coordinate systems, click Mercator (world) > OK.**

**4**  **Zoom to full extent.** The purpose of the Mercator projection is for navigation. Straight lines on the projection are compass bearings for connecting two points. This projection greatly distorts areas near the polar regions and distorts distances along all lines except the equator.

The Mercator projection is a conformal projection, meaning that it preserves small shapes and angular relationships.

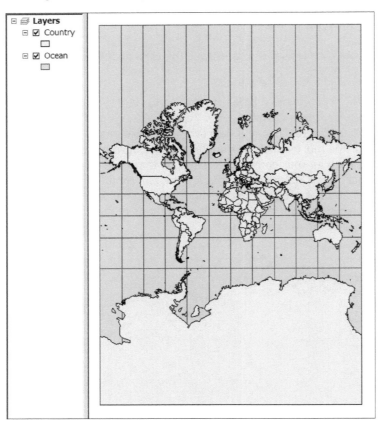

## YOUR TURN

Repeat the steps of the previous exercise, but this time select the Hammer-Aitoff projection in the third step. This projection, shown in the first image, is nearly the opposite of the Mercator. The Hammer-Aitoff is good for use on a world map, being an equal-area projection that preserves area. However, it distorts direction and distance. Repeat the steps again, this time choosing the Robinson projection. As shown in the second image, this projection minimizes distortions of many kinds, striking a balance between conformal and equal-area projections. Save the map document.

**Hammer-Aitoff Projection**

**Robinson Projection**

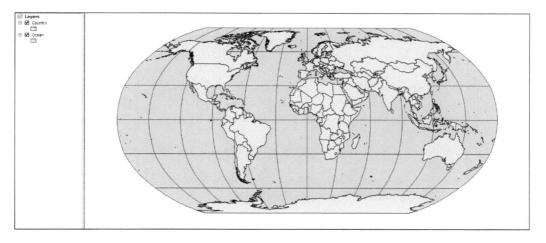

## TUTORIAL 5-3
# Working with US map projections

For most GIS applications it's good to use an equal-area projection of a large area such as the contiguous US states. Then you get accurate areas of polygons and can calculate or use accurate population densities.

# Set projections of the United States

Next, you get some experience with projections commonly used for maps of the continental United States. Some projections are standard for organizations. For example, Albers Equal Area is the standard projection of both the US Geological Survey and the US Census Bureau.

**1**   **Open Tutorial5-3.mxd from the Maps folder.** Initially, the map display is in geographic coordinates. The coordinates will display as decimal degrees.

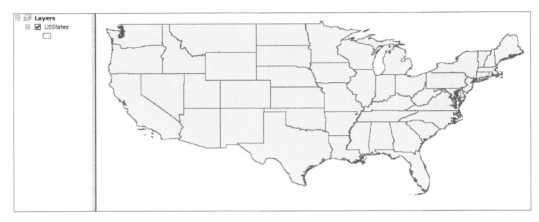

**2**   **Save the map document to the Chapter5 folder.**

**3**   **In the table of contents, right-click Layers and click Properties > Coordinate System tab.**

**4**   **Expand Projected Coordinate Systems > Continental > North America.**

**5**   **Click North America Albers Equal Area Conic, and then click OK and Yes.** Your map is now projected using the Albers Equal Area projection.

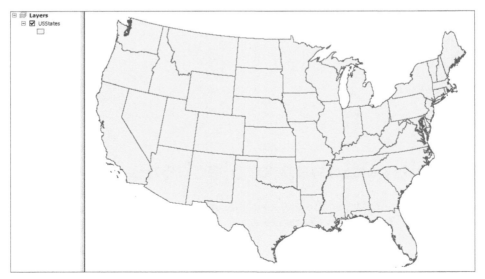

## YOUR TURN

Experiment by applying a few other projections to the US map such as North America Equidistant. As long as you stay in the correct group—Continental, North America—all the projections look similar. The conclusion is that the smaller the part of the world that you need to project, the less distortion. There remains much distortion at the scale of a continent, but much less so than for the entire world. By the time you get to a part of a state, such as Allegheny County, practically no distortion is left, as you'll see next. Save your map document.

## TUTORIAL 5-4

# Working with rectangular coordinate systems

For medium- and large-scale maps it's helpful to have localized projections, tuned for the study area, that have little or no discernible distortion. For this purpose, there are collections of projections. You must check a reference map to determine which zone your study area is in and select that projection for your map.

## State plane coordinate system

The state plane coordinate system is not a projection but a coordinate system dividing the 50 US states, Puerto Rico, and the US Virgin Islands into more than 124 numbered zones, each with its own finely tuned projection. Used mostly by local government agencies such as counties, municipalities, and cities, the state plane coordinate system is for large-scale mapping in the United States. The US Coast and Geodetic Survey developed it in the 1930s to provide a common reference system for surveyors and mapmakers. The first step in using the state plane coordinate system is to look up the correct zone for your area and consequently a specific projection tailored to your study area. Pennsylvania's Allegheny County is in State Plane Zone 3702 (LAMBERT | SOUTHERN ZONE).

# Add projected layers to a map document

As a default, the first map layer that you add to a map document sets the coordinate system and projection for the data frame. If all your map layers have spatial reference data included, you should have no problem combining maps with different coordinate systems or projections. ArcMap will project all map layers to the data frame's projection on the fly. Next, you add a layer with a State Plane Coordinate System projection and then a layer with geographic coordinates.

**1**  **In a new blank map, click the Add Data button, browse to the Data folder, AlleghenyCounty. gdb, and add Munic to the map.** The coordinates appearing in the lower-right corner of the display now appear in state plane units (feet). The origin of these coordinates (0,0) is at the lower-left corner of Pennsylvania.

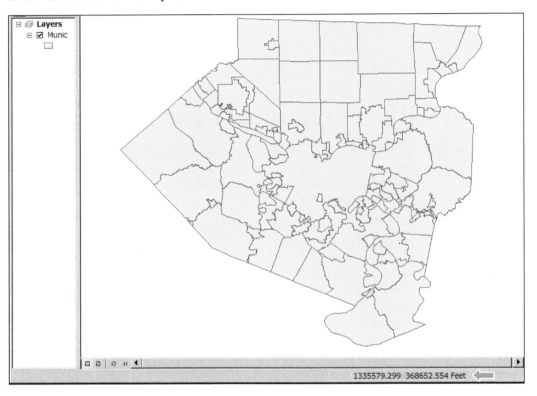

**2**  **Save your map document as** Tutorial5-4.mxd **to the Chapter5 folder.**

**3** In the table of contents, right-click Layers and click Properties > Coordinate System tab. Notice that the data frame coordinates are the same as the Munic layer, NAD_1983_ StatePlane_Pennsylvania_South_FIPS_3702_Feet. This is because it was the first layer added to the new map document. ArcMap will automatically project all additional layers to state plane, regardless of their original coordinates if the added layers have spatial reference data included.

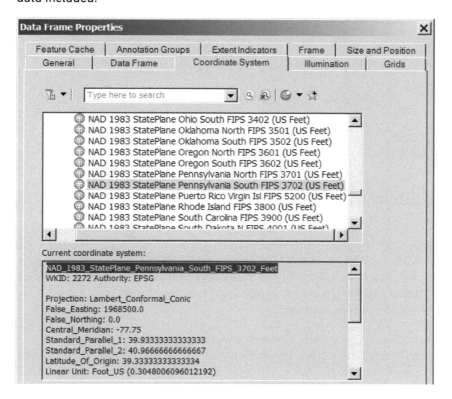

**4** Click OK and change the Munic symbology to a hollow fill, black outline, width 1.15.

**5** Click the Add Data button, browse to AlleghenyCounty.gdb, and add Tracts2010 to the map.

**6** Change the Tracts2010 symbology to a hollow fill, light-gray (20%) outline and move below Munic in the table of contents.

**7** Right-click Munic in the table of contents and click Properties and the Source tab, and note the coordinate system in the Data Source panel. This layer has state plane coordinates, the same as the data frame.

**8** **Repeat step 7 except for Tracts2010.** This layer has geographic coordinates but appears in the data frame using state plane coordinates because you added Munic first and it has state plane coordinates.

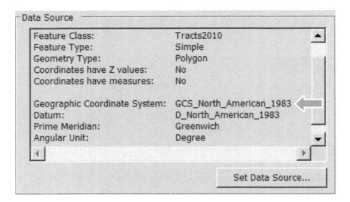

## Change the data frame's coordinate system and projection to UTM

The US military developed the universal transverse Mercator (UTM) rectangular coordinate system in the late 1940s. It includes 60 longitudinal zones defined by meridians that are 6° wide. ArcGIS has UTM projections available for the northern and southern hemisphere of each zone. These zones, like state planes, are good for areas about the size of a state (or smaller) and have the advantage of covering the entire world.

**1** **Start your web browser, go to www.dmap.co.uk/utmworld.htm, and determine the UTM zone for western Pennsylvania.** You should find that western Pennsylvania is in zone 17 north.

**2** **Close your browser.**

**3** **In the table of contents, right-click the Layers data frame and click Properties.**

**4** **Click the General tab and note that the map units are feet.**

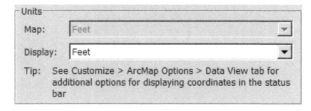

**5** **Click the Coordinate System tab and expand Projected Coordinate Systems > UTM > NAD 1983. Click NAD 1983 UTM Zone 17N.** UTM is a metric system and thus uses meters as the map unit. Your display remains in feet so you need to set the display units to meters.

**6** **Click the General tab and click Meters as the display units > OK.** The coordinate system and map appearance change slightly accordingly. Notice that the coordinates are now in meters.

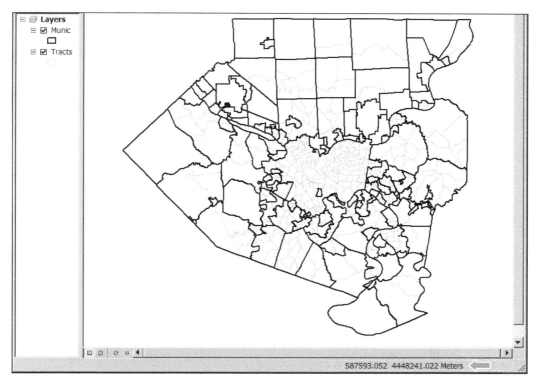

## Assign a coordinate system and projection to a feature class

All GIS layers should have a coordinate system and projection defined (that is, have spatial reference data), but sometimes you will receive a shapefile or other GIS layer that does not have this data and you will need to assign this yourself. Note that the needed spatial reference data is not metadata but data that is part of the map layer. In a shapefile, it is a separate file and has a .prj extension. A common coordinate system for North American files is the Geographic Coordinate System (GCS), North American Datum 1983 (NAD 1983) projection, which is used by organizations such as the US Census Bureau and is the coordinate system of the shapefile that you are about to use.

**1** **In Catalog, navigate through the Data > DataFiles folder, right-click AlleghenyCountyBlockGroups.shp, and click Properties > XY Coordinate System tab.** The coordinate system projection for this shapefile is <Unknown>. Of course you must know the correct coordinate system to assign to the map layer from external information, metadata, or from other sources. If you assign the wrong coordinates, your map will not display correctly in your map document.

**2** **Expand Geographic Coordinate Systems, expand North America > USA and territories, click NAD 1983, and click OK.**

**3**  Repeat step 1 to see that the layer now has its coordinate system data included.

**4**  Add AlleghenyCountyBlockGroups to your map document. You might see a warning that the data is in a coordinate system different from that of the current data but this is OK because it has its native coordinate system assigned. Notice that the block groups overlay nicely with the other map layers that are now projected as UTM rectangular coordinates.

**5**  Remove AlleghenyCountyBlockGroups from the table of contents.

**6**  Save your map document.

## TUTORIAL 5-5
# Learning about vector data formats

This tutorial reviews several file formats commonly found for vector spatial data, other than the file geodatabase covered in chapter 4. Included are Esri shapefiles and coverages as well as computer aided design (CAD) files and XY event files.

## Examine a shapefile

Many spatial data suppliers use the shapefile data format for vector map layers because it is so simple. Shapefiles appeared about the same time that personal computers became popular. A shapefile consists of at least three files: an SHP file, a DBF file, and an SHX file. Each of these files uses the shapefile's name but with the different file type extensions. The SHP file stores the geometry of the features, the DBF file stores the attribute table, and the SHX file stores an index of the spatial geometry for speeding up processing. Next, you examine AlleghenyCountyBlockGroups.shp in more detail.

**1**  Examine AlleghenyCountyBlockGroups.shp in Catalog. It appears as an entry in one line with an icon representing a polygon map layer. ArcCatalog treats the several files as a unit and provides utilities such as renaming the shapefile in one location.

**2  Open File Explorer and navigate to the Data > DataFiles folder.** Now you can see that there are five files for the shapefile, including the projection (.prj) file that you created in tutorial 5-4 when you added a spatial reference for the layer's coordinate system. Shapefiles have three to seven associated files.

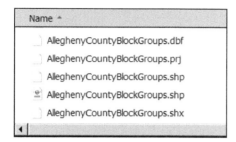

**3  Close File Explorer.**

## Add a coverage to ArcMap

A coverage is a legacy Esri spatial data format from times when personal computers did not even exist. Coverages typically store one or more related feature classes in a folder named for the coverage. For example, in a cadastral (landownership) dataset, it is common for a coverage to store the parcel boundaries as polygons and the parcel lines making up the polygons as arcs (lines). You can add coverage data to ArcMap and use it for analysis and presentation, but you cannot edit coverage data with ArcMap. When browsing data within File Explorer, coverages appear as folders containing several files. In the following image, you can see the four coverages in the EastLiberty folder with the contents of the Building coverage appearing in the right side of File Explorer. The Building coverage has 18 files.

1    In a new blank map, click the Add Data button, browse through the Data > Pittsburgh > EastLiberty folder, and double-click Building.

2    **Click the Polygon layer icon, click Add, and OK.** A coverage behaves like any other vector layer in ArcMap. It has the same appearance and has an attribute table.

3    **Save the map document as** Tutorial5-5.mxd **to the Chapter5 folder.**

---

### YOUR TURN

Add Curbs Arc and Parcel Polygon coverages to your map for the East Liberty neighborhood. Change the symbols to your liking and examine the attribute table for the Parcel layer.

---

## Convert a coverage to a feature class

If you need to edit the attribute tables or geometry of a coverage, you must export it first as a shapefile or file geodatabase feature class.

1    **In the table of contents, right-click the Building Polygon layer and click Data > Export Data.**

2    **Browse to Chapter5.gdb in the Chapter5 folder of MyExercises and save the output feature class as** Building.

3    **Click OK > Yes to add the exported data to the map as a layer.** Now you could edit the polygons in the Building feature class and add the missing spatial reference for the layer. You learn about editing features in chapter 7.

4    **Save the map document.**

## Add a CAD drawing file

Many organizations have CAD (computer aided design) drawings that you can display in ArcMap in their native format. ArcMap can add CAD files in one of two formats: as native AutoCAD (.dwg) or as Drawing Exchange Files (.dxf) that most CAD software can create. When viewed in Catalog, a CAD dataset appears with a light-blue icon. An AutoCAD file is much like a coverage in that it has different kinds of vector features in the same file. You can see CampusMap.dwg in ArcCatalog in the figure.

**1**  In a new blank map, click the Add Data button, browse through the Data > CMUCampus folder, click the CampusMap.dwg icon, click Add, and click OK to the spatial warning message. The map of the Carnegie Mellon University campus appears in ArcMap as a layer group that contains many feature types, including lines, polygons, and text. This map is from the Facilities Management Department and has spatial reference data.

**2**  In the table of contents, if not already expanded, expand the CampusMap.dwg Group Layer and turn off all layers except CampusMap.dwg Polygon. This shows academic and residential buildings on CMU's campus.

**3**  Right-click the CampusMap.dwg Polygon layer and click Properties > Drawing Layers tab. Notice that you can turn layers on and off. You can also symbolize the layers using the Symbology tab.

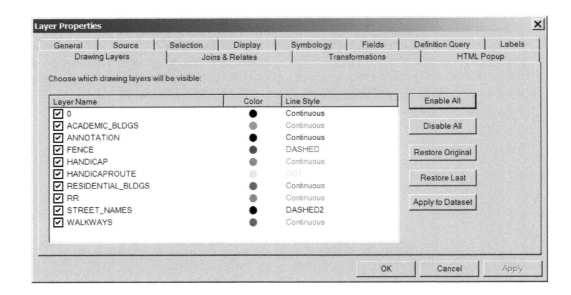

**4**  Close the Layer Properties window and remove CampusMap.dwg Group Layer from the table of contents.

## Export GIS features to CAD

Sometimes you may need to deliver GIS features to a person working with CAD software. You can export features to AutoCAD (.dwg) or Drawing Exchange Files (.dxf) formats, which can then be opened by most commercial CAD applications.

**1**  Click the Add Data button, browse through the Data folder to 3DAnalyst.gdb, click Topo, and click Add. This is two-dimensional topography.

**2** In the table of contents, right-click the Topo layer and click Data > Export to CAD.

**3** Type or make selections as shown, with Topo for Input Features, DWG_R2010 for Output Type, Topo.dwg for Output File, and Append to Existing Files (Optional) checked. Click OK.

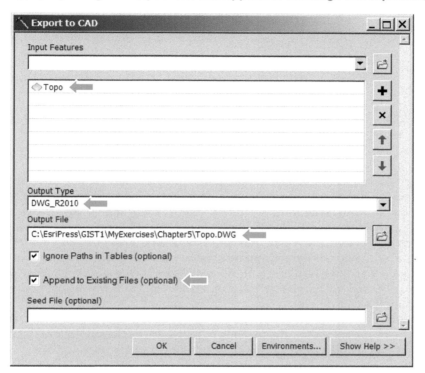

The CAD drawing is automatically added to the table of contents. You can also open Topo.dwg in any computer application that supports the CAD .dwg format.

**4**    Remove the Topo and CAD drawing layers from the table of contents.

**5**    Save your map document as Tutorial5-5.mxd to the Chapter5 folder.

## TUTORIAL 5-6
# Exploring raster basemaps from Esri web services

Whereas vector maps are discrete—consisting of points, lines connecting points, and polygons made up of lines—raster maps are continuous, like photographs, and use many of the same file formats as images on computers, including Joint Photographic Experts Group (.jpg) and tagged image file (.tif) formats. All raster maps are rectangular, consisting of rows and columns of cells known as pixels. Each pixel has an associated coordinate and an attribute value such as altitude for elevation. Raster maps do not store each pixel's location explicitly but rather store data such as the coordinates of the northwest corner of the map, cell size (assuming square pixels), and the number of rows and columns, from which a computer algorithm can calculate the coordinates of any cell. A key aspect of raster maps from your point of view is that they are very large files. So although you may store some important basemap raster files on your computer, these kinds of maps are perhaps best obtained as map services available for display on your computer but stored elsewhere.

## Add a basemap layer

You can add several different basemap layers to a map document as services in addition to map layers from your computer or local area network. ArcGIS has built-in access to Esri servers for this purpose.

**1**    Open a new blank map and save it as Tutorial5-6.mxd to the Chapter5 folder.

**2**    Browse to the Data folder > Chicago.gdb, and add tl_2010_17031_cousub10 to the map.

**3**    Symbolize the County Subdivisions layer with a hollow fill and a yellow outline, width 1.15.

**4** Click **File > Add Data > Add Basemap > Imagery with Labels > Add.** It takes a few moments for the map image to download to ArcMap. If your basemap does not appear, right-click County subdivisions and click Zoom To Layer. The map image that displays is from an Esri server instead of your computer's hard disk. You can now more clearly see Lake Michigan.

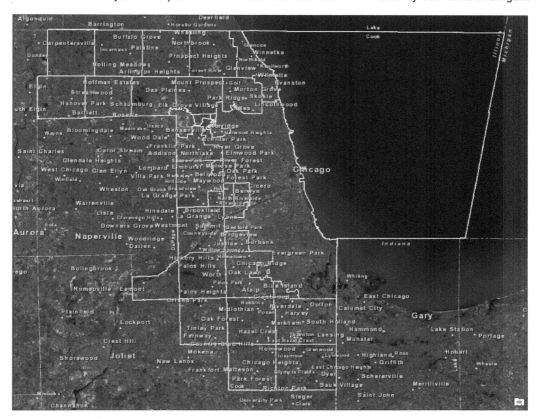

**5**  Zoom to an area in Chicago near Lake Michigan and Millennium Park.

## YOUR TURN

Remove the current basemap from the table of contents and repeat step 3 for another basemap. Save your map document.

## TUTORIAL 5-7
# Downloading raster maps from the USGS

Next, you'll download an elevation raster from the US Geological Survey that can be added in ArcMap. The raster map in this exercise is a digital elevation model (DEM) that can be turned into a shaded relief map, meaning that it includes shadows from an artificial sun in the sky. In chapter 11, you will learn how to convert a DEM into a shaded relief model, as shown in the image.

**Launch the USGS National Map Viewer and zoom to Pittsburgh**

**1** In a web browser, go to http://viewer.nationalmap.gov/basic/.

**2**  **Zoom to western Pennsylvania and Pittsburgh where the three rivers meet until you see a scale of approximately 1:36,000 like the image.** Note that the screen captures and layers might look different from what you see on your computer.

Scale approximately 1:36,000

## Download the DEM

**1** On the Datasets tab, check the box for Elevation Products (3DEP).

**2** Check the box for 1/9 arc-second DEM and click Find Products.

**3** From the Available Products list, click Download next to the item USGS NED ned19_ n40x50_w080x25_pa_southwest_2006 1/9 arc-second 2010 15 x 15 minute IMG.

**4** On the dialog box, choose to save the file to your computer.

## Extract and organize

**1** Using File Explorer, locate your file from the download location.

**2** Extract the file to your Chapter5 folder.

**3** Use File Explorer to view these files. Close File Explorer when you are finished.

| | | |
|---|---|---|
| NED_DataDictionary | Internet Shortcut | 1 KB |
| ned19_n40x50_w080x00_pa_southwest_2006.dbf | DBF File | 1 KB |
| ned19_n40x50_w080x00_pa_southwest_2006 | Disc Image File | 257,599 KB |
| ned19_n40x50_w080x00_pa_southwest_2006.img.aux | XML Document | 3 KB |
| ned19_n40x50_w080x00_pa_southwest_2006.prj | PRJ File | 1 KB |
| ned19_n40x50_w080x00_pa_southwest_2006.rrd | RRD File | 87,443 KB |
| ned19_n40x50_w080x00_pa_southwest_2006.sbn | SBN File | 1 KB |
| ned19_n40x50_w080x00_pa_southwest_2006 | Adobe Illustrator ... | 1 KB |
| ned19_n40x50_w080x00_pa_southwest_2006 | AutoCAD Shape S... | 1 KB |
| ned19_n40x50_w080x00_pa_southwest_2006 | AutoCAD Compil... | 1 KB |
| ned19_n40x50_w080x00_pa_southwest_2006_meta | HTML Document | 24 KB |
| ned19_n40x50_w080x00_pa_southwest_2006_meta | Text Document | 13 KB |
| ned19_n40x50_w080x00_pa_southwest_2006_meta | XML Document | 14 KB |
| ned19_n40x50_w080x00_pa_southwest_2006_thumb | JPEG image | 15 KB |
| ned19_readme | Adobe Acrobat D... | 27 KB |
| readme | Adobe Acrobat D... | 113 KB |

## Add the raster image to the map

**1** In ArcMap, create a new blank map and save it as Tutorial5-7.mxd to the Chapter5 folder.

**2** Click the Add Data button; browse to Data, Pittsburgh folders, and City.gdb; click Neighborhoods and click Add. Symbolize neighborhoods with a hollow fill.

**3** Click the Add data button, navigate to MyExercises, Chapter5, and view the extracted raster files. Click the .img file, click Add, and then Yes to build raster pyramids. The DEM raster image will automatically be classified and added to the table of contents.

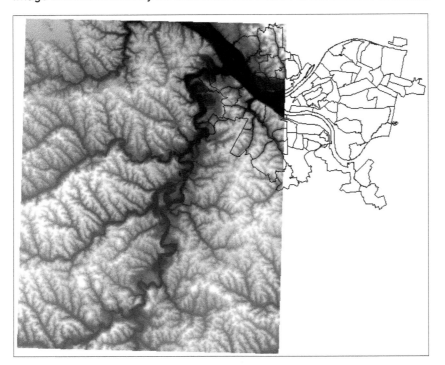

# 6

# Geoprocessing

Generally you will need to use geoprocessing tools to build study areas in a GIS. In this chapter, you'll learn how to extract a subset of spatial features from a map using attribute or spatial queries, aggregate polygons to larger polygons, and append two or more layers into a single layer. Often it is necessary to string several geoprocessing tools and steps together to build the desired product, so you'll also learn how to create macros using the ArcGIS® ModelBuilder™ application to bundle two or more steps together into a single package.

Learning objectives

- Use attribute and spatial queries to extract features
- Clip features
- Dissolve features
- Merge features
- Intersect layers
- Union layers
- Automate geoprocessing with ModelBuilder

## TUTORIAL 6-1
# Extracting features for a study area

New York City is the most populated US city with more than 8 million people; it has a population density of more than 27,000 per square mile. With 195 neighborhoods in five boroughs, it is difficult to analyze or visualize the entire city at once. If city officials want to study one neighborhood in detail, it can be done using ArcMap geoprocessing tools. In this tutorial you'll use two methods—select by attributes and select by location—to create study area features for one neighborhood.

## Open a map document

**1** **Start ArcMap and open Tutorial6-1.mxd from the Maps folder.** The map document opens with a map zoomed to Manhattan, showing surrounding New York City metropolitan area boroughs, neighborhoods, and Manhattan streets.

**2** **Click Bookmarks > Manhattan.**

**3** **Save the map document to the Chapter6 folder of MyExercises.**

## Use Select By Attributes to select features

Here you use the ArcMap Select By Attributes tool to create a study area for the Lower East Side neighborhood extracted from the Neighborhoods layer.

**1** **On the Menu bar, click Selection > Select By Attributes.**

**2** **Select Neighborhoods for the Layer.**

**3** **In the Fields box, double-click Name.**

**4** **Click the = button.**

**5** Click the Get Unique Values button. Then, in the Unique Values box, double-click 'Lower East Side.'

**6** Click Apply and click OK.

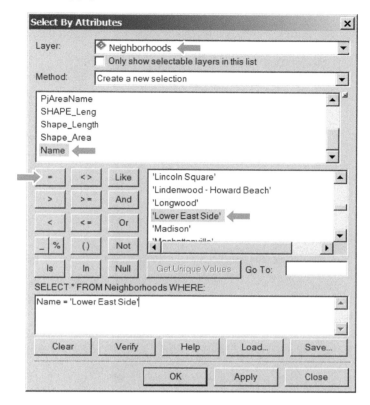

# Zoom to and export selected features

**1** Click Selection > Zoom to Selected Features. ArcMap zooms to the Lower East Side neighborhood in the Manhattan borough.

**2**  In the table of contents, right-click the Neighborhoods layer and click Data > Export Data.

**3**  Save the output feature class as LowerEastSide in **Chapter6.gdb to the Chapter6 folder of MyExercises.**

**4**  Click OK > Yes to add the layer to the map.

**5**  Symbolize the LowerEastSide layer as a hollow fill, black outline, width 2.

**6**  Remove the Boroughs and Neighborhoods layers. Your map now contains a new feature class containing only the Lower East Side neighborhood boundary.

## Use Select By Location to select features

In the following steps, you use the Select By Location tool to select Manhattan streets that intersect the Lower East Side neighborhood only. After selecting the streets, you will create a new line feature class from them.

**1**  Click Selection > Select By Location.

**2**  Make selections as shown in the image, with select features from for Selection method, ManhattanStreets for Target layer(s), LowerEastSide for Source layer, and intersect the source layer feature for Spatial selection method for target layer feature(s). This specification selects only streets in the Lower East Side.

**3**  Click Apply and click Close.

**4**  In the table of contents, right-click the Manhattan Streets layer and click Data > Export Data.

**5**  Save the output feature class as LowerEastSideStreets in **Chapter6.gdb** and add the layer to your map document.

**6**  Symbolize the LowerEastSideStreets layer as light-gray lines and zoom to this layer.

**7**  In the table of contents, move LowerEastSideStreets below LowerEastSide and remove the Manhattan Streets layer. You now have a study area for the Lower East Side. Using Select By Location produces streets that "dangle" past the neighborhood outline. In tutorial 6-2, you'll use Clip to create streets that stop at the neighborhood outline.

## YOUR TURN

Use Select By Location to create a new feature class of LowerEastSideFacilities in Chapter6. gdb. Use "are within the source layer feature" as the selection method. Remove the original Facilities layer and symbolize the new layer using Circle 2, Mars Red, size 7. Save your map document.

## TUTORIAL 6-2

# Clipping features

Next, you will use the Clip geoprocessing tool to cleanly "cut off" the Lower East Side street segments using the Lower East Side feature class. Once this is done, the streets will have no dangling lines. Note that for geocoding tabular address data with streets, you should use the streets version with dangles because ArcMap interpolates house numbers using the starting and ending house numbers of street segments. Streets that ArcMap clips will have the original starting and ending house numbers but shortened lengths. This will introduce location errors beyond those inherent in the approximate streets. Use the clipped streets in a study area for display purposes only.

# Clip streets

**1** **Save your map document as** Tutorial6-2.mxd **to the Chapter6 folder.**

**2** **Click Geoprocessing > Clip.**

**3** **In the Clip window, select LowerEastSideStreets as Input Features.**

**4** **Select LowerEastSide for Clip Features.**

**5** **Save the Output Feature Class as** LowerEastSideStreetsClipped **in Chapter6.gdb.**

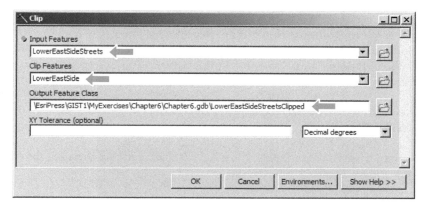

**6** **Click OK and wait while the streets are clipped.**

**7** **Turn on the LowerEastSideStreetsClipped layer and turn off the LowerEastSideStreets layer.** As seen in the map, the streets in the LowerEastSideClipped layer do not cross the Lower East Side neighborhood boundary.

**8** Save your map document.

# TUTORIAL 6-3
# Dissolving features

You can create administrative or other types of boundaries by dissolving polygons in a feature class that share common attribute values. For each group of original polygons, dissolving retains the outer boundary lines of the group but erases interior lines. The New York City Fire Department, like most fire departments around the world, is organized in a quasi-military fashion with companies, battalions, and divisions. In this tutorial you use the Dissolve tool to dissolve New York City fire company polygons to create fire battalions and fire divisions.

## Open a map document

**1** Open Tutorial6-3.mxd from the Maps folder. Tutorial6-3.mxd contains a map of the New York City fire companies and boroughs.

**2** Save the map document to the Chapter6 folder.

# Examine the dissolve attribute

**1** In the table of contents, right-click FireCompanies and click Open Attribute Table. The FireBN and FireDiv attributes are used to aggregate into fire battalion and fire division polygons. The field Pop2010 shows the population in each fire company and can be aggregated during the dissolve process.

| | OBJECTID * | Shape * | FireCo Type | FireCo Num | FireBN | FireDiv | Shape_Length | Shape_Area | Pop2010 |
|---|---|---|---|---|---|---|---|---|---|
| | 1 | Polygon | E | 153 | 21 | 8 | 26830.204525 | 29213947.52536 | 22221 |
| | 2 | Polygon | E | 206 | 28 | 11 | 30274.034187 | 15192741.33212 | 3338 |
| | 3 | Polygon | E | 214 | 37 | 15 | 12069.312309 | 8852559.675643 | 17747 |
| | 4 | Polygon | E | 216 | 35 | 11 | 13253.213817 | 9463049.329291 | 17847 |
| | 5 | Polygon | E | 217 | 57 | 11 | 17584.543354 | 9738320.205785 | 21123 |
| | 6 | Polygon | E | 222 | 37 | 15 | 14060.23304 | 10061437.89181 | 18536 |

**2** Close the table.

# Dissolve fire battalions

**1** Click Geoprocessing > Dissolve.

**2** In the Dissolve window, select FireCompanies as the Input Features.

**3** Save the Output Feature Class as FireBattalions in Chapter6.gdb.

**4** Select FireBN as the Dissolve field.

**5** Select POP2010 for the Statistics Field(s). Note: Ignore the red x warning.

**6** Click the Statistic Type drop-down and choose SUM. This is an optional setting. When the dissolve runs, ArcGIS sums the values in the POP2010 field for each group of polygons with the same FireBN value. In other words, it aggregates the population up to the new polygon feature of fire battalions.

**7** Verify your selections with the following image as shown, with FireCompanies for Input Features, FireBattalions for Output Feature Class, FireBN for Dissolve_Field(s), and SUM for Statistic Type.

**8**  Click OK and wait while the polygons dissolve.

**9**  Symbolize the new layer with a hollow fill, an Ultra Blue outline color, and an outline width of 1.15.

**10** Zoom to Lower Manhattan and use the Identify tool to view the attribute information for fire battalion number 4 in the **Lower East Side neighborhood.** The SUM_POP2010 value is the aggregate of the dissolve process derived from the POP2010 values of the FireCompanies layer, at 152,002.

## YOUR TURN

Use the Dissolve tool and field FireDiv to create fire divisions. Sum field POP2010 while dissolving. Symbolize the new layer with no fill, a Tuscan Red outline color, and an outline width of 1.15. Save your map document.

## TUTORIAL 6-4
# Merging features

Sometimes it is necessary to merge two or more separate but adjacent layers into a single layer. For example, you may want to build a water layer for an environmental study that includes water layers from several adjacent counties. New York City is made up of five boroughs, each of which is also a county. The Bronx is also Bronx County, Brooklyn is Kings County, Manhattan is New York County, Queens is Queens County, and Staten Island is Richmond County. Here you will merge water layers for New York City's counties.

## Open a map document

**1**  **Open Tutorial6-4.mxd from the Maps folder.** Tutorial6-4.mxd contains a map of the New York City area counties' water features, shown as separate layers.

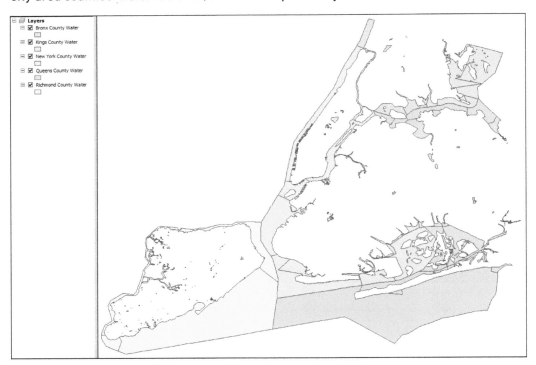

**2**  Save the map document to the Chapter6 folder.

## Merge several feature layers into one feature class

**1**  Click Geoprocessing > Merge.

**2**  In the Merge window, select all five layers for the New York area water polygons for Input Datasets.

**3**  For the Output Feature Class save NYCWater in Chapter6.gdb.

**4** Verify that the Merge settings match the image.

**5** Click OK.

**6** **In the table of contents, remove the individual county water layers.** The NYCWater layer now contains water boundaries for all five counties around New York City.

**7** Save your map document.

## YOUR TURN

From the NYC.gdb file geodatabase in the Data folder add the following layers: BronxWaterfrontParks, BrooklynWaterfrontParks, ManhattanWaterfrontParks, QueensWaterfrontParks, and StatenIslandWaterfrontParks. Use the Merge tool to create one feature class named NYCWaterfrontParks in Chapter6.gdb. Save your map document.

## TUTORIAL 6-5
# Intersecting layers

The Intersect tool creates a new feature class combining all the features and attributes of two input, overlaying feature classes. For example, an emergency preparedness official might like to know the name of the fire company that each street crosses over (or intersects). ArcGIS can provide such information using the Intersect tool. Intersect excludes any parts of the two or more input layers that do not overlay each other.

## Open a map document

**1** **Open Tutorial6-5.mxd from the Maps folder.** The Tutorial6-5.mxd file opens with a map of the New York City fire companies and Manhattan streets.

**2** **Click Bookmarks > Manhattan.**

**3** **Save the map document to the Chapter6 folder.**

# Open tables

Next, you'll open the table of both feature classes to see their attributes.

**1** From the table of contents, right-click the ManhattanStreets layer, click Open Attribute Table, and scroll to the right. There is no data about the fire companies in this file.

| | LLo_Hyphen | LHi_Hyphen | RLo_Hyphen | RHi_Hyphen | FromLeft | ToLeft | FromRight | ToRight | Join_ID | SHAPE_Leng | Shape_Length |
|---|---|---|---|---|---|---|---|---|---|---|---|
| ► | 2 | 20 | 1 | 21 | 2 | 20 | 1 | 21 | 2800801000000 | 218.580099 | 218.580033 |
| | 2901 | 2999 | 2900 | 2998 | 2901 | 2999 | 2900 | 2998 | 2712001000000 | 250.556354 | 250.55629 |
| | | | | | 0 | 0 | 0 | 0 | 2460801020000 | 163.243858 | 163.244044 |
| | 5487 | 5495 | 5486 | 5494 | 5487 | 5495 | 5486 | 5494 | 2107201000000 | 39.486989 | 39.487197 |
| | | | | | 0 | 0 | 0 | 0 | 24922509000000X | 39.486989 | 39.487197 |
| | | | | | 0 | 0 | 0 | 0 | 2990502000000 | 385.46756 | 385.467739 |
| | | | | | 0 | 0 | 0 | 0 | 2990502000000 | 946.666951 | 946.666832 |
| | | | | | 0 | 0 | 0 | 0 | 2990502000000 | 363.57388 | 363.573819 |

ManhattanStreets

I◄ ◄   1 ► ►I    (0 out of 27315 Selected)

**2** Open the FireCompanies table. Examine the attributes of this table. Fields of interest are the fire company, battalion, and division numbers and the type of fire company; for example, ladder company (L), engine company (E), or fire squad (Q).

FireCompanies

| OBJECTID * | Shape * | FireCoType | FireCoNum | FireBN | FireDiv | Pop2010 | Shape_Length | Shape_Area |
|---|---|---|---|---|---|---|---|---|
| 1 | Polygon | E | 153 | 21 | 8 | 22221 | 26830.204525 | 29213947.525366 |
| 2 | Polygon | E | 206 | 28 | 11 | 3338 | 30274.034187 | 15192741.332126 |
| 3 | Polygon | E | 214 | 37 | 15 | 17747 | 12069.312309 | 8852559.675643 |
| 4 | Polygon | E | 216 | 35 | 11 | 17847 | 13253.213817 | 9463049.329291 |
| 5 | Polygon | E | 217 | 57 | 11 | 21123 | 17584.543354 | 9738320.205785 |
| 6 | Polygon | E | 222 | 37 | 15 | 18536 | 14060.23304 | 10061437.891819 |
| 7 | Polygon | E | 227 | 44 | 15 | 9861 | 12410.32117 | 5331327.239706 |
| 8 | Polygon | E | 247 | 40 | 8 | 40804 | 24622.662513 | 21978391.299848 |
| 9 | Polygon | E | 248 | 41 | 15 | 27383 | 21875.193282 | 19248897.323501 |

I◄ ◄   1 ► ►I    (0 out of 348 Selected)

**3** Close both tables.

# Intersect feature layers

**1** Click Geoprocessing > Intersect.

**2** From the Input Features drop- down list, select ManhattanStreets and Fire Companies one at a time.

**3** Save the Output Feature Class as ManhattanStreetsFireCompanies in Chapter6.gdb.

**4** Select Line for Output Type.

**5** Verify that the Intersect settings match the image.

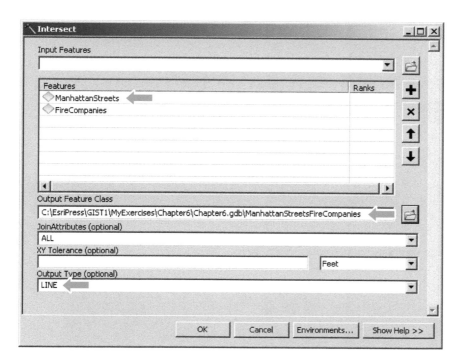

**6** Click OK. The output added to your map will be Manhattan streets that intersect the fire company polygons.

## Examine the intersection table

1    In the table of contents, right-click the ManhattanStreetsFireCompanies layer and click
Open Attribute Table. Each street now has data about the fire company that it intersects.

ManhattanStreetsFireCompanies

| FromRight | ToRight | Join_ID | SHAPE_Leng | FID_FireCompanies | FireCoType | FireCoNum | FireBN | FireDiv | Pop2010 | Shape_Length |
|---|---|---|---|---|---|---|---|---|---|---|
| 0 | 0 | 3043401000000 | 417.327294 | 253 | E | 202 | 32 | 11 | 6541 | 417.327204 |
| 0 | 0 | 3631237010000 | 100.750667 | 177 | L | 106 | 28 | 11 | 17494 | 100.750685 |
| 0 | 0 | 1076601040000 | 100.476275 | 253 | E | 202 | 32 | 11 | 6541 | 100.476218 |
| 0 | 0 | 1059911010800 | 58.832465 | 311 | E | 35 | 12 | 3 | 20049 | 58.832551 |
| 0 | 0 | 1059911010800 | 63.555795 | 311 | E | 35 | 12 | 3 | 20049 | 63.555955 |
| 0 | 0 | 1043301000000 | 791.627747 | 44 | E | 40 | 9 | 3 | 27500 | 791.627341 |
| 0 | 0 | 1043301000000 | 94.674368 | 42 | E | 34 | 7 | 1 | 3631 | 94.674398 |
| 0 | 0 | 1043301000000 | 1960.798869 | 42 | E | 34 | 7 | 1 | 3631 | 1960.79885 |

| ◄  ◄    1  ►  ►|    ▦▦  (0 out of 29627 Selected) |

2    Close the table and save your map document.

## TUTORIAL 6-6

# Unioning layers

The Union tool combines the geometry and attributes of two input polygon layers to generate
a new output polygon layer. In this example, you will use the Union tool to combine neighborhoods in New York City with Manhattan (New York County) census tracts. The output of the
union is a new feature layer of smaller polygons, each with combined boundaries and attributes of both census tracts and neighborhoods. Union keeps all features of the input layers,
even if they do not overlap.

## Open a map document

1    Open Tutorial6-6.mxd from the Maps folder. The Tutorial6-6.mxd file opens showing a map
of the New York City neighborhoods and Manhattan census tracts 2010.

**2** Click Bookmarks > Manhattan.

**3** Save the map document to the Chapter6 folder.

## Open tables

The union function creates a new feature class that combines the attribute tables of two feature classes and their polygon features. Next, you will open the table of both feature classes to see their attributes.

**1** From the table of contents, right-click the ManhattanTracts layer, click Open Attribute Table, and scroll to the right. There is no data about the neighborhoods in this file.

| AWATER10 | INTPTLAT10 | INTPTLON10 | Id | Id2 | TOT_POP | MALE | FEMALE | AGE_UNDER_1 | AGE_18_64 | AGE_65_UP |
|---|---|---|---|---|---|---|---|---|---|---|
| 0 | +40.8519392 | -073.9342905 | 36061027100 | 36061027100 | 8196 | 4016 | 4180 | 1663 | 5507 | 1026 |
| 407508 | +40.7650139 | -073.9998117 | 36061012900 | 36061012900 | 6038 | 3432 | 2606 | 501 | 5185 | 352 |
| 0 | +40.7752486 | -073.9475520 | 36061014401 | 36061014401 | 4864 | 2161 | 2703 | 753 | 3245 | 866 |
| 1125670 | +40.7523775 | -074.0083669 | 36061009900 | 36061009900 | 1945 | 1002 | 943 | 192 | 1694 | 59 |
| 0 | +40.7771918 | -073.9521610 | 36061014601 | 36061014601 | 4274 | 1962 | 2312 | 384 | 3583 | 307 |
| 0 | +40.7813494 | -073.9491245 | 36061015400 | 36061015400 | 13749 | 6127 | 7622 | 1594 | 10395 | 1760 |
| 0 | +40.7842784 | -073.9521037 | 36061015801 | 36061015801 | 5650 | 2545 | 3105 | 894 | 3875 | 881 |
| 0 | +40.7835686 | -073.9475024 | 36061015601 | 36061015601 | 5345 | 2249 | 3096 | 498 | 4538 | 309 |

(0 out of 288 Selected)

**2**  Open the Neighborhoods table. Of course, there are no tract attributes in this table.

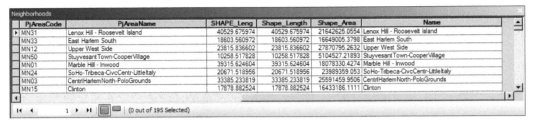

**3**  Close both tables.

## Select and export Manhattan neighborhoods

Before performing the union, select only Manhattan neighborhoods. Otherwise the union output will be all boroughs.

**1**  Click Selection > Select By Location.

**2**  Type or make selections as shown, with Neighborhoods for Target layer(s), ManhattanTracts for Source layer, and "have their centroid in the source layer feature" for Spatial selection method.

**3**  Click OK.

**4** In the table of contents, right-click the Neighborhoods layer and click Data > Export Data.

**5** Save the output feature class as ManhattanNeighborhoods in Chapter6.gdb.

**6** Click OK, and then click Yes to add the layer to the map.

**7** Symbolize the ManhattanNeighborhoods layer as a hollow fill, black outline, width 1.5.

**8** Remove the Neighborhoods layer and zoom to lower Manhattan. Turn the ManhattanNeighborhoods layer off and on to see which polygons match the tracts boundaries. Some tracts extend beyond the neighborhood boundaries.

**9** Click Bookmarks > Manhattan.

## Union feature classes

**1** Click Geoprocessing > Union.

**2** Select ManhattanTracts and ManhattanNeighborhoods one at a time for Input Features.

**3** Save the Output Feature Class as ManhattanTractsNeighborhoods in Chapter6.gdb.

**4** Verify that the Union settings match the image, with ALL for JoinAttributes.

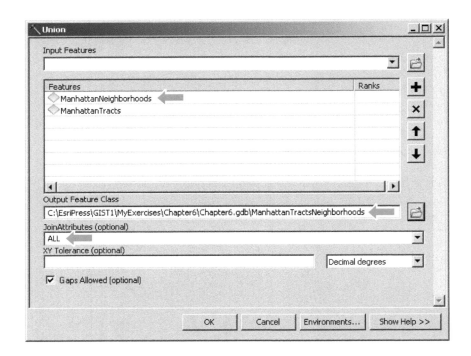

**5**   **Click OK.** The output added to your map contains many small polygons with both census tract and neighborhood data attached to each polygon. Note that the population for each new small polygon is incorrect. The Union tool merely joins the layers together and does not apportion the data across the smaller polygons.

## Select tracts with neighborhoods

Many polygons from the union do not include neighborhood names because tracts and neighborhoods do not overlap. The final exercises exclude these polygons from the map document. Here you'll use a query to do so.

**1**   In the table of contents, right-click ManhattanTractsNeighborhoods and click Properties > Definition Query tab.

**2**   Click the Query Builder button and make selections as shown in the image, with WHERE set to BoroCode > 0.

**3** **Click OK > OK.** The result will be only ManhattanTractsNeighborhoods with neighborhood values.

**4**  In the table of contents, right-click ManhattanTractsNeighborhoods and click Open Attribute Table. There are still some census tracts without neighborhood names but these are parks and cemeteries. Notice the fields for census tracts and neighborhoods.

**5**  Close the table and save the map document.

## TUTORIAL 6-7
# Automating geoprocessing using ModelBuilder

Spatial data processing often requires several steps and geoprocessing tools to produce desired results. ModelBuilder is an application in ArcGIS for creating macros—custom programs that document and automate geoprocessing workflows. After you build a model, you can run it once, or save it and run it again using different input parameters. In this tutorial, you'll build a model with several steps for dissolving census tracts to make neighborhoods for a city within a county. Before you go to work using ModelBuilder, it is helpful to examine the inputs and outputs, and then the finished model that you build.

The starting map document, shown in the image, has all municipalities (cities) and census tracts in Allegheny County, Pennsylvania, as downloaded from the US Census Bureau's TIGER basemaps. In this tutorial, you'll create neighborhoods for Pittsburgh in the center of the map.

| OID | STFID | HOOD |
|---|---|---|
| 0 | 42003220400 | Allegheny Center |
| 1 | 42003220100 | Allegheny West |
| 2 | 42003180300 | Allentown |
| 3 | 42003160300 | Arlington |
| 4 | 42003160400 | Arlington Heights |
| 5 | 42003202300 | Banksville |
| 6 | 42003050900 | Bedford Dwellings |
| 7 | 42003191600 | Beechview |
| 8 | 42003192000 | Beechview |
| 9 | 42003180900 | Beltzhoover |
| 10 | 42003080200 | Bloomfield |
| 11 | 42003080400 | Bloomfield |
| 12 | 42003080600 | Bloomfield |
| 13 | 42003080900 | Bloomfield |
| 14 | 42003090300 | Bloomfield |
| 15 | 42003010300 | Bluff |

The user must supply a crosswalk table that lists the tracts that define neighborhoods in the city. In this case, each Pittsburgh neighborhood is made up of one or more tracts, as seen in the partial crosswalk table listing in this table.

The output is the dissolved set of neighborhoods. You can see the tracts that ModelBuilder dissolved for each neighborhood as interior black lines for the red neighborhoods.

When you run the model, a form opens asking you to supply parameters—all the elements in the model that the user needs to change for the run. With the model, it is possible to create dissolved polygons for any subset of a polygon basemap. See the following figure for a description of this model's parameters. This is the user interface for the model that has documentation and parameters that the user can change.

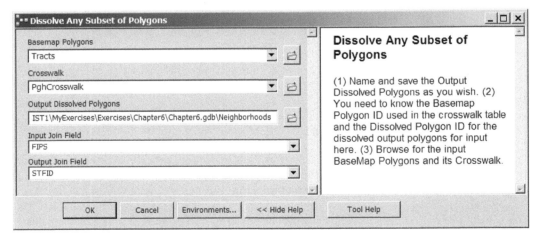

The workflow model diagram that you build is shown in the image. Earlier in this chapter you ran geoprocessing steps interactively from the ArcMap main menu. For models, however, you access the same functionality using the ArcToolbox window and tools. Each tool becomes a process (the yellow boxes seen in the model in the following figure) with blue inputs and green outputs in a model diagram.

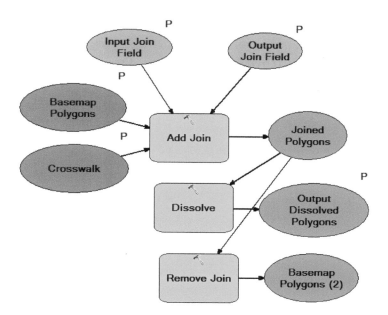

ModelBuilder shows input and output elements with black arrow lines that go from and to processes. Each element with a *P* near its upper right is a parameter. Input Join Field and Output Join Field are variables that store input parameter values for further processing (an explanation is given later in this tutorial).

The first step of the model is to join the crosswalk table to the basemap polygons. The user can supply any two consistent inputs to Add Join that have matching polygon IDs. The output, Joined Polygons, has only polygons included in the crosswalk table. In the case that you will run, the crosswalk is for Pittsburgh tracts, so only Pittsburgh tracts are output from the larger county basemap.

Next, the Dissolve process uses the crosswalk data to carry out dissolving, resulting in Output Dissolved Polygons. Finally, the model removes the join so that you can rerun the model with the same or different initial inputs. Otherwise, an error would result, indicating that the join already exists.

## Open a map document

1    **Open Tutorial6-7.mxd from the Maps folder.** The map of Allegheny County opens displaying TIGER file census tract and municipality polygons. Municipalities is just for reference, while Tracts is an input for dissolving. The other input, the crosswalk table PghCrosswalk, is also available in the map document.

2    **In the Table of Contents window, click the List by Drawing Order button.**

3    **Save the map document to the Chapter6 folder of the MyExercises folder.**

## Set geoprocessing options

**1**  Click Geoprocessing > Geoprocessing Options.

**2**  If it is not already selected, make sure that "Overwrite the outputs of geoprocessing operations" is checked. With this option on, you can rerun the model repeatedly without having to delete model outputs first, which saves time when debugging and getting your model to work properly.

**3**  Click OK.

## Create a new model

**1**  Click Windows > Catalog.

**2**  Expand Home—MyExercises\Chapter6 in the folder/file tree.

**3**  Right-click Home—MyExercises\Chapter6, click New and Toolbox, and rename the new toolbox Chapter6.tbx.

**4**  Right-click Chapter6.tbx and click New, Model. ArcMap opens the Model window that you use to create your model.

## Join the crosswalk table to the layer to dissolve

Next, you browse through system tools to find the Dissolve tool. When you pursue model building on your own, you will need to systematically browse through all the tools available to get ideas and see what is possible. When you find a tool and want to learn about it, right-click it and click Help.

**1**  Close Catalog and click Windows > Search.

**2**  Click Tools from the Search window. The result is a listing and links to the ArcGIS classification of tools available for use directly or as elements in models.

**3**  Click the Data Management tools link. Here you see the first page of many pages with data management toolsets.

**4**  Scroll down and click Joins.

**5**  Drag the Add Join (Data Management) tool to your model. Close the Search window.

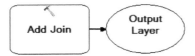

**6** Double-click the Add Join process in your model and make selections using the drop-down list in each field as shown in the image, with Tracts for Layer Name or Table View, FIPS for Input Join Field, PghCrosswalk for Join Table, and STFID for Output Join Field. Ignore the information message symbol for the Input Join Field. Be sure to clear the Keep All Target Features (optional) check box. With this option off, the only features kept in the output are those in the crosswalk table, which will be Pittsburgh census tracts. This saves a Clip tool step.

**7** Click OK, and resize and reposition model elements as shown in the image.

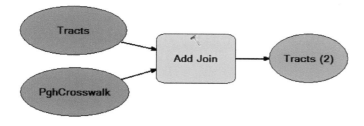

**8** Click the model's Save button.

## Run the partial model

ArcMap appends table names to attribute names in joined data; for example, HOOD in PghCrosswalk becomes PghCrosswalk.HOOD. Next, you'll run the Add Join process. Then the appended attribute, PghCrosswalk.HOOD, is available in a list of attributes for you to use in configuring additional processes in subsequent steps. If you didn't run the Add Join process now, you'd have to know about and type the appended name on your own.

**1** Right-click Add Join in your model and click Run. As the process runs, a report window opens for the task and its status.

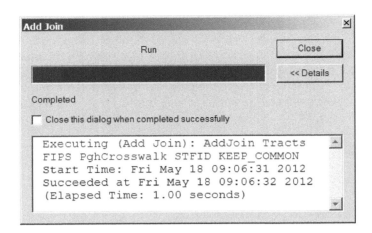

**2**  Click Close. ArcGIS adds shadows to the process and its output to indicate that they have been run. Note that if you make an error and need to rerun the model, first you must click Model on the Model window's main menu, and then click Validate Entire Model. This resets all processes to the unrun state.

## Dissolve tracts

**1**  Click Windows > Search, type dissolve in the search text box, and press Enter.

**2**  Drag the Dissolve (Data Management) tool below the Add Join process in your model. Close the Search window.

**3**  Click the Connect button ᵉ on the Model window's Standard toolbar, click Tracts (2) output from the Add Join process in the model, click the Dissolve process, and click Input Features in the resulting context menu.

**4**  Click the model's Select button ▸ . You always should click this button after using another button on the Standard tool bar. Otherwise, the next time you click in the Model window, you will get an undesired action. The Select button's action is usually acceptable.

**5**  Double-click the Dissolve process in your model and make selections using the drop-down list in each remaining field as shown in the image —with Tracts(2) for Input Features, Neighborhoods for Output Feature Class, and PGHCrosswalk.HOOD for Dissolve_Field(s)— but do not click OK.

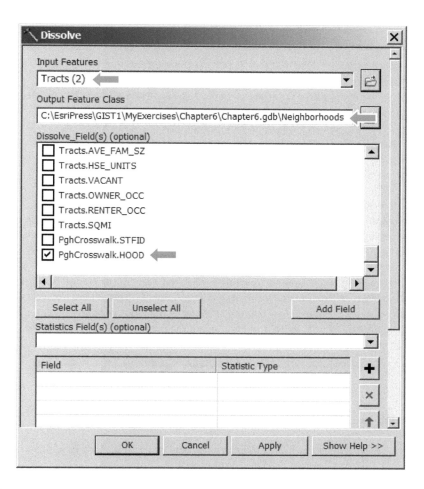

**6**  Select Tracts.POP2000 in the Statistics Field(s) and ignore the warning.

**7**  Click in the Statistic Type cell to the right of Tracts.Pop2000, click the resulting drop-down arrow, and select SUM.

**8**  Repeat steps 6 and 7 for two additional attributes, Tracts.WHITE and Tracts.BLACK, using SUM for both. At this point, you should have Tracts.Pop2000, Tracts.WHITE, and Tracts.BLACK in the lower panel of the Dissolve tool window, all with the SUM Statistic Type.

**9**  Click OK, right-click the Neighborhoods output of the Dissolve process, click Add To Display, and save your model.

**10**  Right-click the Dissolve process, click Run, and close the resulting window when the model has finished running.

## YOUR TURN

The basic model is almost complete. The last step is to have the model remove the join in the Tracts output of the Add Join process so that the user can run the model again without manually doing so. Otherwise, the Add Join process would fail because a join already exists. Search for and add the Remove Join (Data Management) tool to the model as the last process. Use the output of Add Join, Tracts (2), as its input (double-click the Remove Join process, click "Layer Name or Table View," select the output of the Add Join process, and select PghCrosswalk for the Join). The Remove Join tool automatically identifies PghCrosswalk as the join to. Do not add the output of Remove Join to the display.

Run the Remove Join process. Save your model. Symbolize Output Joined Polygons (the neighborhoods) with a hollow fill and a red outline with width 3, and then move Tracts to the top of the table of contents and compare Tracts and the new Neighborhoods. You should see the output display that is at the start of this tutorial on ModelBuilder.

Reset the model so that you can run it again by clicking Model, Validate Entire Model. Run the entire model by clicking Model > Run Entire Model. Again, symbolize Output Joined Polygons with a hollow fill and red size 3 outline, move Tracts to the top of the table of contents, and compare Tracts and the new Neighborhoods. You should see the output display that is at the start of this tutorial on ModelBuilder.

## Generalize element labels

Your model is capable of being a general tool for dissolving any polygons. As a first step to making the model general, you change several element labels.

1   Right-click the Tracts input to Add Join element, click Rename, type Basemap Polygons, and click OK.

2   Similarly, change labels of other elements as shown in the image, with Crosswalk as another element, and Joined Polygons, Output Dissolved Polygons, and Basemap Polygons(2) for the results.

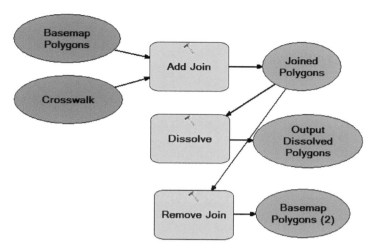

**3**  Click the model's Save button.

## Add model parameters

Currently, your model is "hardwired" with inputs and outputs fixed in processes. Next, you will make several element parameters that users can change without modifying the model itself. Instead, users will type or make selections in a form when opening the model.

**1**  Right-click Basemap Polygons and click Model Parameter. A P appears above and to the right of the element, indicating ArcGIS will ask the user to browse for an input map layer.

**2**  Similarly, make Crosswalk and Output Dissolved Polygons model parameters.

## Add variables to the model

To be general, the Add Join process needs to get two of its inputs from the user, the Input Join Field and the Output Join Field. You can make these inputs parameters, but first you'll need to create variables to store them in the model.

**1**  Right-click the Add Join process. Click Make Variable > From Parameter > Input Join Field. ArcGIS creates the variable for you.

**2**  Click anywhere in the white area of the model to deselect elements. Move the new variable above the top left of the Add Join process and make its element a bit wider so that its entire label displays.

**3**  Make Input Join Field a model parameter.

**4**  Repeat steps 1 through 3, except make the variable for the Output Join Field of the Add Join process. Move the new element above and to the right of the Add Join process.

# Add labels for documentation

Labels can help document the model. You will add a model title and some notes about the variables.

**1**  If necessary, select all model elements and make some room at the top for a label.

**2**  Right-click the white area at the top, click **Create Label**, double-click the resulting **Label**, **and type** Model to Dissolve a Subset of Basemap Polygons.

**3**  Right-click the new label, click **Display Properties**, click in the cell to the right of Font, click the resulting builder button, change the font to Bold size 14, click OK, and close the Display Properties window.

**4**  Right-click the Input Join Field element, click **Create Label**, click anywhere in the model white area, drag the new label above and to the left of the element, double-click it, and **type** You must examine the Basemap Polygons attribute table and Crosswalk table and note their field names that share the same polygon IDs or names for use as parameter values on opening the model.

**5**  Right-click the new label, click **Display Properties**, click the cell to the right of Text Justification, click Left instead of Center, and close the Properties window.

**6**  Break this label up into several lines by placing your cursor after words to create new lines, pressing the Shift key, and pressing Enter.

**7**  Click the model's Save button.

### Model to Dissolve a Subset of Basemap Polygons

You must examine the Basemap attribute table
and Crosswalk table and note their field names
that share the same polygon IDs or names for
use as parameter values when opening the model.

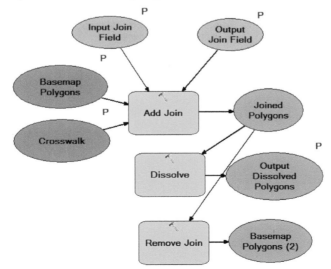

# Add a model name and description for documentation

You can add help documentation to the form that will open on running the model, in model properties.

**1**  **Click Model on the model's main menu and click Model Properties.**

**2**  **Type** DissolvePolygons **for Name (no spaces are allowed) and** Dissolve Any Subset of Polygons **for Label.**

**3**  **Type as much as you want of the following help message in the Description text box:** Dissolve any subset of basemap polygons given a crosswalk table for the subset.

1.  Name and save the Output Dissolved Polygons as you wish.
2.  You need to know the Basemap Polygon ID used in the crosswalk table and the Dissolved Polygon ID for the dissolved output polygons for input here.
3.  Browse for the input Basemap Polygons and its Crosswalk.

**4**  **Select the Store relative path names check box.**

**5**  **Click OK.**

**6**  **Click the model's Save button and close the Model window.**

# Open and run the finished model

**1**  **Remove Neighborhoods from the table of contents in your map document.**

**2**  **In Catalog, right-click the Dissolve Any Subset of Polygons model, click Open, and close the warning at the top of the resulting form.**

**3**  **Click the Show Help button at the bottom of the form.** You do not need to change any of these input parameters.

**4**  **Click OK to run the model.** The model runs, adding the dissolved neighborhoods to your map display.

**5**  **Save your map document and close ArcMap.**

## ASSIGNMENT 6-1

# Build a study region for Colorado counties

In this assignment, you build a study area for two rapidly growing counties in Colorado: Denver and Jefferson Counties. You create new feature classes for an urban area study using point and polygon layers downloaded from the US Census website. Because we want to study two counties, you'll need to use select by location, merge, and clip to create the study area features.

## Get set up

- Rename the folder **\EsriPress\GIST1\MyAssignments\Chapter6\Assignment6-1YourName\** to your name or student ID. Store all files that you produce for this assignment in this folder. Create a new map document named **Assignment6-1YourName.mxd** with relative paths.
- Create a new file geodatabase named **Assignment6-1YourName.gdb**. Store new features that you produce for this assignment in this file geodatabase.

## Build the map

Add the following to your map document:

- \EsriPress\GIST1\Data\UnitedStates.gdb\COCounties—polygon features of Colorado counties.
- \EsriPress\GIST1\Data\UnitedStates.gdb\COStreets1—line features of Jefferson County streets.
- \EsriPress\GIST1\Data\UnitedStates.gdb\COStreets2—line features of Denver County streets.
- \EsriPress\GIST1\Data\UnitedStates.gdb\COUrban1—US Census 2010 urban area features for Jefferson County.
- \EsriPress\GIST1\Data\UnitedStates.gdb\COUrban2—US Census 2010 urban area features for Denver County.
- \EsriPress\GIST1\Data\UnitedStates.gdb\COWater1—US Census 2010 named water features for Jefferson County.
- \EsriPress\GIST1\Data\UnitedStates.gdb\COWater2—US Census 2010 named water features for Denver County.
- \EsriPress\GIST1\Data\UnitedStates.gdb\USCities—point features of US cities.

# Process the data

Perform select by location or geoprocessing operations to create the following study area features in your file geodatabase. Remove the original features when finished with the geoprocessing.

- **StudyAreaCounties**—one feature class of Jefferson and Denver counties only
- **StudyUrbanAreas**—one feature class combining urban areas for both Jefferson and Denver counties
- **StudyAreaCities**—point layer showing populations of cities in the study urban area only. Add and calculate a field for the population change between 2000 and 2007.
- **StudyAreaStreets**—one feature class combining streets also clipped to the study urban area
- **StudyAreaWater**—one feature class combining water features that intersect the study urban area

# Requirements

- Label the study area counties and cities using the NAME fields with halo masks, display the cities as graduated points using the POP2007 field with a quantile classification and 5 classes, and display the water and streets as ground features. Decide how best to display the study urban area polygons. Select Wadsworth street in the study area. **Hint:** Use FE_NAME=Wadsworth. Show it as a separate layer.
- Create an 8.5-by-11-inch map layout zoomed to the study urban area that includes a title, legend, scale bar, and other items you think necessary. Export the layout as **Assignment6-1YourName.jpg**.

# Analyze the map

Create a Word document named **Assignment6-1YourName.docx**. Answer the following questions using the attribute tables and spatial queries, and insert your layout image below your answers:

1. What is the total 2007 population of the cities in the Jefferson County study areas?
2. What is the total 2007 population of the city in Denver County?
3. What city had the largest population increase between 2000 and 2007?
4. What city had the largest population decrease between 2000 and 2007?
5. Name the cities that are within two miles of Wadsworth roads in the study urban area. **Hint**: Select all streets, boulevards, parkways, and so forth, named Wadsworth.

# Dissolve property parcels to create a zoning map

In this assignment, you dissolve a parcel map to create a zoning map that highlights a proposed commercial development in what is now a residential area. A commercial company wants to apply for a zoning variance so that it can use the land in residential parcels with zoning code R5 (residential dwelling with five units) for a commercial purpose. The zoning department wants the map for use in a PowerPoint presentation at a public hearing on the proposal.

## Get set up

- Rename the folder **\EsriPress\GIST1\MyAssignments\Chapter6\Assignment6-2YourName\** to your name or student ID. Store all files that you produce for this assignment in this folder.
- Create a new map document named **Assignment6-2YourName.mxd** with relative paths and data frame map units in feet.
- Create a new file geodatabase named **Assignment6-2YourName.gdb**. Store new features that you produce for this assignment in this file geodatabase.

## Build the map

Add the following coverages to your map document:

- \EsriPress\GIST1\Data\Pittsburgh\EastLiberty\EastLib—polygon coverage for the East Liberty neighborhood boundary.
- \EsriPress\GIST1\Data\Pittsburgh\EastLiberty\Parcel—polygon coverage for land parcels in the East Liberty neighborhood of Pittsburgh. Attributes include: ZON_CODE, an attribute with the following zoning code values: A = development, C = commercial, M = industrial, R = residential, S = special; TAX_AREA_A, TAX_BLDG_A, and TAX_LAND—attributes with annual property tax values.
- \EsriPress\GIST1\Data\Pittsburgh\EastLiberty\Curbs—polyline coverage that has street curbs and annotation with street names.

## Requirements

- Export the Parcels coverage to your file geodatabase and add it to your map document. Remove the original Parcels coverage.

- Select parcels with ZON_CODE = R5. These are parcels that could be converted for a commercial purpose. Add these parcels as a new layer renamed **Proposed Commercial Parcels** shown with red color fill and black outline.
- Create an aggregate-level zoning code by adding a new field to the parcels attribute table that has just the first character of the full zoning code. Name the new field **ZONE** with text data type and length 1. Using the Field Calculator on the new field, click the String Option button, click the Left( ) function, type to yield **Left([ZON_CODE],1)**, and click OK. The left function extracts the number of characters entered—1 in this case—starting on the left of the input field, ZON_CODE.
- Dissolve the parcel's features using the Dissolve function, using your new field (ZONE) as the dissolve field, and adding SUM statistics for the three tax fields (TAX_LAND_A, TAX_BLDG_A, TAX_AREA). Click in the Statistics Type cells to select SUM. Save the output feature class named **Zoning** in your file geodatabase.
- Add the new Zoning feature class to your map document, as well as the curbs arcs and curbs annotation, and East Liberty outline. Use the Unique Values option of the categories method of classification for symbolizing the ZONE field. Use muted colors with no outlines for the various ZONE values, including Lilac Dust for A (Development), Rose Quartz for C (Commercial), Gray 30% for M (Industrial), Yucca Yellow for R (Residential), and Blue Gray Dust for S (Special). Label each zoning code with its full name. Add zoning features again as a light-gray, hollow fill outline.
- Create an 8.5-by-11-inch portrait layout zoomed to scale 1:9,000 including a title, legend, map scale in feet, and other map elements that you think necessary.
- Export your map layout as **Assignment6-2YourName.jpg**.

## Create a PowerPoint document

Create a PowerPoint presentation named **Assignment6-2YourName.pptx** that includes the following:

- Title slide, including your name
- Slide with a table of the summed Tax values for each zoning code
- Slide with your map layout image

## ASSIGNMENT 6-3
# Build a model to create a fishnet map layer for a study area

The following image is the fishnet map layer that you will create with several steps saved in a ModelBuilder model. It consists of uniform, square grid cells saved in a polygon map layer, and as an option includes an additional layer of centroid points for each cell. Both the cells and centroids are useful for spatial analysis; for example, for displaying counts of structural fires by grid cell. The centroids allow you to display a second attribute using size-graduated point markers or a color ramp, while a choropleth map displays the first variable. You can create cell-level data from point data, such as residential fire incidents or crimes locations, using spatial overlay as done in chapter 2.

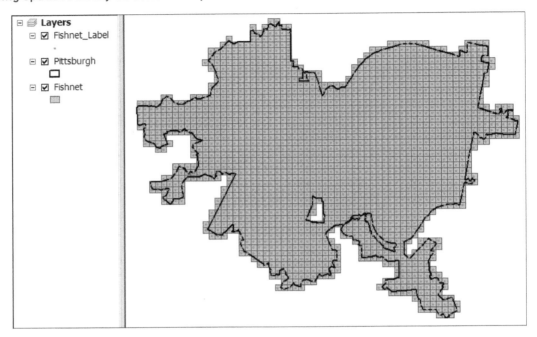

## Get started

First, rename your assignment folder, create a map document, create a file geodatabase, and create a toolbox and model.

- Rename the folder **\EsriPress\GIST1\MyAssignments\Chapter6\Assignment6-3YourName\** to your name or student ID. Store all files that you produce for this assignment in this folder.
- Create a new map document, **Assignment6-3YourName.mxd**, with relative paths.
- Create a new file geodatabase, **Assignment6-3YourName.gdb**. Save all the map layers and tables you create in this file geodatabase.
- Add a new toolbox, **Assignment6-3YourName.tbx**, with a model named **FishnetStudyArea**.
- Add \EsriPress\GIST1\Data\Pittsburgh\City.gdb\Pittsburgh to your map document, symbolized with a hollow fill.

## Requirements

The model diagram for the finished model that you create is in the figure. The input map layer is any polygon layer that defines the study area. It could be a boundary, such as for Pittsburgh, census tracts, or any other set of polygons. The Create Fishnet tool (see tutorial 2-7) has several parameters for constructing the grid cell map layer. It needs the cell size (while the tool has inputs for both width and height, you will almost always want square grid cells, so width will equal height), the number of rows and columns in the extent, and the extent coordinates of the output. The Create Fishnet tool has provision to import the needed map extent coordinates from the input study area polygons, so that part is easy. To calculate the number of rows and columns of the fishnet, you must have projected coordinates for the input layer (the Pittsburgh layer has projection state plane in feet, so this case meets that requirement) and you must look up its map extent coordinates in its Source properties as seen in the image as Top, Bottom, Left, and Right. You can get these extent values by right-clicking Pittsburgh in the Table of Contents and opening its property sheet to the Source tab. Then use the Excel workbook, \EsriPress\GIST1\Data\DataFiles\FishnetCalculations.xlsx, to compute the number of rows and columns for any cell size. For the case of a 1,000-foot cell size, you need 52 rows and 64 columns.

| | A | B | C | D | E | F |
|---|---|---|---|---|---|---|
| 2 | | | | | | |
| 3 | Cell size | 1,000 | Can be changed to whatever you wish | | | |
| 4 | | | | | | |
| 5 | **Number of rows** | | | | | |
| 6 | Top | 433,417 | | | | |
| 7 | Bottom | 381,929 | Divide Height by cell size | | | |
| 8 | Height | 51,487 | number cells = | 51.48744 | | |
| 9 | | | **rounded up =** | | 52 | |
| 10 | **Number of columns** | | | | | |
| 11 | Right | 1,379,786 | Divide Width by cell size | | | |
| 12 | Left | 1,315,935 | number cells = | 63.851 | | |
| 13 | Width | 63,851 | **rounded up =** | | 64 | |
| 14 | | | | | | |

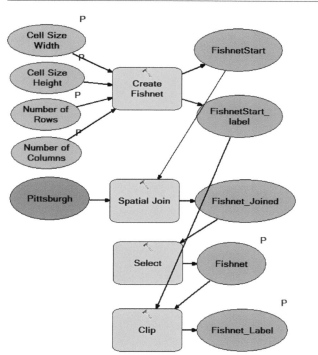

Notes on the Create Fishnet process follow:

- Create the three variables: **Cell Size**, **Number Rows**, and **Number Columns**. Select Double for the data type of CellSize and Long for the other two variables' data type (which is integer).
- For the Create Fishnet process, name the study area polygon layer **(Pittsburgh) StudyArea** and select it for the Template Extent input.
- For the Create Fishnet's Cell Size Width and Height inputs, use the drop-down lists and select Cell Size (the name of your model variable). For the Number of Rows and Number of Columns inputs, use the drop-down lists and select your corresponding variables.
- For the Geometry, choose POLYGON.

The fishnet process creates a rectangular map layer, but what is needed is a grid cell map limited to the shape of the study area. It is best to leave grid cells whole squares, with some overhanging the study area, rather than clipping them. The model needs two steps to accomplish this. First, you must assign study area attributes to grid cells with the Spatial Join process. Then all grid cells in the interior or crossing the boundary of the study area will have non-null attributes while remaining cells will have null values. Then you can use the Select process with the criterion:

**NAME > "**

where there are two single quotes after the greater than sign, signifying a null value. The criterion selects all cells having non-null values.

Finally, you can clip the label points using the finished fishnet. Making the final outputs (**Fishnet_Finished** and **Fishnet_LabelPointsFinished**) parameters allows the user to rename and save them wherever desired.

# 7

# Digitizing

Using existing map layers such as streets for guides, in this chapter you'll learn how to create and edit spatial data. You'll learn how to create and digitize new vector features and add their attribute data. Many advanced digitizing tools are available to assist you, making the work easier and more precise. You also will adjust vector data spatially to make it align with a basemap layer. For example, you will learn how to import and adjust a computer-aided design (CAD) drawing for use in a map layer.

Learning objectives

- Digitize polygon, line, and point features
- Add attribute data for vector features
- Learn advanced editing tools
- Spatially adjust features

## TUTORIAL 7-1
# Digitizing polygon features

You'll create a new polygon feature class, and then add features to it using heads-up digitizing (with your head up, looking at your computer screen).

## Open a map document

**1**   **Start ArcMap and open Tutorial7-1.mxd from the Maps folder.** The Tutorial7-1 map document shows a map of the Middle Hill neighborhood of Pittsburgh. You use the Commercial Properties and Street Centerlines layers as references for digitizing commercial zones.

**2**   **Save your map document to the Chapter7 folder of MyExercises.**

## Create a new polygon feature class

**1**   **Open the Catalog window.**

**2**   **In the Catalog tree, browse to MyExercises, the Chapter7 folder.**

**3**   **Right-click Chapter7.gdb and click New > Feature Class.**

**4**   **In the Name field of the New Feature Class window, type** CommercialZones.

**5**   **For Type, select Polygon Features and click Next.**

**6**   **Expand Projected Coordinate Systems > State Plane > NAD 1983 (US Feet), click NAD 1983 StatePlane Pennsylvania South FIPS 3702 (US Feet), and click Next three times.**

**7**   **Type** ZoneNumber **as a new field, select Short Integer as the Data Type, and click Finish.**

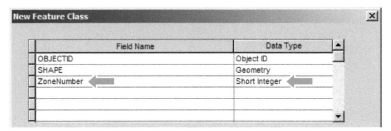

The result is a new polygon feature class added to Chapter7.gdb and the table of contents. Although CommercialZones appears in the table of contents, of course, nothing displays for it because at this point there are no features in this new map layer. Next, you'll digitize new features, starting with practice polygons.

**8**   Close the Catalog window.

# Start editing with the Editor toolbar

**1**   Click Customize > Toolbars > Editor. The Editor toolbar appears. You can move it or dock it anywhere in ArcMap. Dock it on top of the ArcMap window below the Standard toolbar.

**2**   On the Editor toolbar, click Editor > Start Editing.

**3**   Click CommercialZones as the layer to edit and click OK.

**4**   From the Editor toolbar, click the Create Features button. The Create Features and Construction Tools panels appear on the right of the map. You can adjust these panels by dragging the boundary between them.

**5**   Click CommercialZones in the Create Features panel, and then click Polygon in the Construction Tools panel.

**6** Close the Create Features panel.

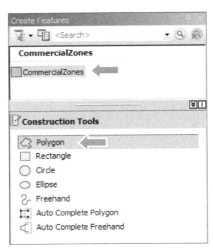

## Practice digitizing a polygon

**1** In the table of contents, click the List By Selection button and make CommercialZones the only selectable layer.

**2** On the Editor toolbar, click the Straight Segment button ⟋.

**3** Position the crosshair cursor anywhere on the map outside the neighborhood boundary and click to place a vertex.

**4** Move your mouse and click a series of vertices one at a time to form a polygon (but do not double-click!). If you click near an existing feature, you'll find that your new vertices snap to existing vertices of other features. You'll learn later how to turn this behavior on and off.

**5**  Double-click to place the last vertex.

## Move a polygon

**1**  On the Editor toolbar, click the Edit tool ▶.

**2**  Press and hold the mouse button anywhere inside your new polygon.

**3**  Drag the polygon a small distance.

## Delete a polygon

**1**  Click anywhere inside your new polygon.

**2**  Press the Delete key on the keyboard.

> ### YOUR TURN
> Practice creating new polygons using the Polygon, Rectangle, Circle, and Ellipse tools from the Create Features, Construction Tools panel. Delete your practice polygons when finished.

# Edit polygon vertex points

Next, you'll learn how to work with vertices. You move, add, and delete vertices from a polygon.

**1**   **Click Bookmarks > Erin Street.** Your map zooms to a few city blocks surrounding Erin Street.

**2**   **Click the Create Features button and from the Construction Tools panel, click Polygon.**

**3**   **Click the Straight Segment button and draw a new polygon feature as shown in the image, snapping to street centerline intersections of Davenport, Webster, Erin, and Wylie.**

**4**   **On the Editor toolbar, click the Edit button.**

**5**   **Double-click the new polygon.** Grab handles—small squares—appear on the polygon at its vertex locations.

Next, you'll see that you can edit the shape of a feature by moving a vertex.

**6**   **Position the cursor over the lower-right vertex point.**

**7**  **Drag the vertex to the street intersections of Wylie and Trent.** The polygon's shape changes correspondingly.

**8**  Drag the upper-right vertex to the intersection of Webster and Trent Streets.

**9**  Click anywhere on the map or polygon to confirm the new shape.

## Add and move vertex points

Next, you'll practice editing digitized polygons and learn how to add, delete, and move vertices.

**1**  **Double-click inside the polygon.** Grab handles appear on the polygon and the small Edit Vertices toolbar appears.

**2**  **On the Edit Vertices toolbar, click the Add Vertex button.**

**3**  **Move the mouse along Wylie Street between the two existing vertices and click.** This adds a new vertex at the location of the cursor. Now you can move the new vertex to change the polygon's shape.

**4**  **Position the cursor over the new vertex, and then drag the vertex to the intersection of Wylie and Erin streets.**

**5** Click anywhere on the map to confirm the new shape.

> ## YOUR TURN
> Practice adding and moving vertex points for the intersections of Webster and Erin streets and Webster and Seal Streets.

## Delete vertex points

**1** Double-click inside the new polygon.

**2** Click the Delete Vertex button ⌐.

**3** Place your cursor over the vertex point between Webster and Erin and click.

**4** Click anywhere on the map to confirm the new shape.

## YOUR TURN

Practice changing the shape of the new polygon by moving, adding, and deleting vertices. When finished, delete the polygon.

## Digitize commercial zones

**1** Click **Bookmarks > LaPlace.** This shows the commercial block centroids in this city block. The image shown has the polygon drawn that you are about to roughly digitize.

**2** In the **Construction tools panel, click Polygon.**

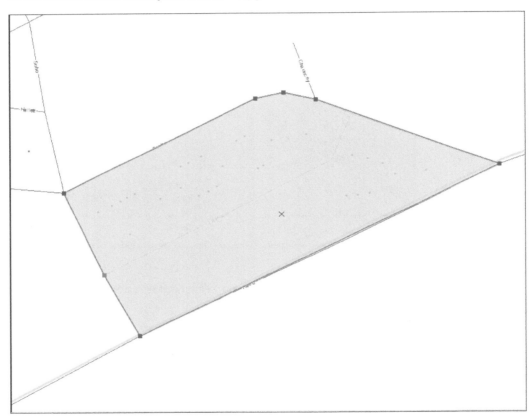

**3** On the **Editor** toolbar, click the **Straight Segment** tool and digitize the LaPlace street polygon by clicking one vertex at a time and double-clicking to finish. Wherever possible, use street centerlines as a guide for digitizing your polygon.

## YOUR TURN

Zoom in to a part of your new polygon and use the Add, Delete, and Move Vertex tools to refine the polygon's shape. Use the Pan tool on the Tools toolbar to move around your polygon's boundary and eventually refine all of it. You need to alternate between the Edit tool, confirming a change, and the Pan tool. Click the Full Extent button and then zoom in to a cluster of commercial points to digitize another polygon. Repeat until you have digitized all polygons seen in the following image. When you complete the final polygon, click Editor and save your edits.

## Edit feature attribute data

Now that you have digitized the commercial polygons, you'll assign zone numbers to them.

**1** Open the CommercialZones attribute table.

**2** Click in the first cell of the ZoneNumber field, type 1, and press Enter.

**3** In sequential order, continue numbering the remaining cells in the ZoneNumber field.

**4** Click Editor > Stop Editing and click Yes to save your edits.

**5** Close the attribute table.

## Label commercial zones

**1** Turn off the Commercial Properties layer.

**2** In the table of contents, right-click the CommercialZones layer and click Properties.

**3** Select the Labels tab and type or make selections as shown, with "Label features in this layer" checked and ZoneNumber as Label Field and a symbol font of Arial 12 pt. bold .

**4** Click OK. Your label numbers may not match those in the following image, depending on your order of digitizing.

**5**  Save the map document.

## TUTORIAL 7-2
# Digitizing line features

City planners often work with transportation experts to decide bus route and stop locations. Existing street centerlines can be used as a base layer to digitize bus route locations. In this exercise, you create the bus routes that are in the Middle Hill neighborhood of Pittsburgh.

## Create a line feature class for bus routes

**1**  Open Tutorial7-2.mxd from the Maps folder and save it to the Chapter7 folder.

**2**  Click Windows > Catalog.

**3**  In the Catalog tree, browse to Chapter7.gdb in the Chapter7 folder.

**4**  Right-click Chapter7.gdb and click New > Feature Class.

**5** In the Name field, type BusRoutes and select Line Features for Type.

**6** Click Next and NAD_1983_StatePlane_Pennsylvania_South_FIPS_3702_Feet as the Current coordinate system.

**7** Click Next three times.

**8** Type ROUTE as a new field, select Text as the Data Type, click Finish, and close the Catalog window. The new bus route feature class will be added to your table of contents, and you are now ready to digitize the routes.

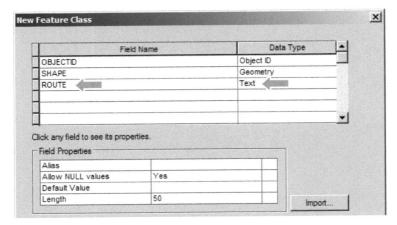

## Symbolize the bus route features

**1** In the table of contents, click BusRoutes's line symbol to open the Symbol Selector window.

**2** Change the bus routes color to Ultra Blue, width 1.15.

# Prepare area for digitizing and start editing

**1**  Click Bookmarks > Centre Avenue Route to zoom to area as shown in the image.

**2**  On the Editor toolbar, click Editor > Start Editing.

**3**  Click BusRoutes > OK.

**4**  On the Editor toolbar, click the Create Features button.

**5**  In the Create Features panel, click BusRoutes > Line in the Construction Tools panel.

## Set endpoint and vertex snapping

Here you'll turn on only endpoint and vertex snapping so you snap to only the endpoint of the street segment. Snapping makes digitizing the bus routes faster.

**1**  Click Customize > Toolbars > Snapping.

**2**  Deselect the point and edge buttons so only the endpoint and vertex buttons are on.

## Digitize by snapping to features

**1**  On the Editor toolbar, click the Straight Segment tool.

**2**  Click the endpoint of the leftmost Centre Avenue street centerline to choose the route's starting point.

**3** Move the cursor to the next street endpoint (Centre and Covet) and click.

**4** Continue snapping to street intersections along the bus route shown in the image and double-click to finish the route at the intersection of Centre and Reed Streets.

## Enter bus route number

It is often easiest to enter the data for a feature as you digitize it. Here you'll enter the bus route number for the route that you just digitized.

**1** In the table of contents, right-click BusRoutes and click Open Attribute Table.

**2** **Type** 81A|81B|84A|84C **for ROUTE.**

**3** **Close the BusRoutes attribute table.**

## YOUR TURN

Digitize bus routes for Bedford/Erin, Webster, Wyle, and Kirkpatrick Streets. Use spatial bookmarks to zoom to the route locations. Enter the route numbers and label as shown in the figure. Be sure to digitize one line for each route. When finished, stop editing and save the edits to your bus routes. Save the map document.

## TUTORIAL 7-3
# Digitizing point features

In addition to bus routes, point features for bus stops are needed for planning purposes. In this exercise, you'll digitize bus stop locations as points.

## Create a point feature class for bus stops

**1**  **Save your map document as** Tutorial7-3.mxd **to the Chapter7 folder.**

**2**  **Click Windows > Catalog.**

**3**  **In the Catalog tree, browse to Chapter7.gdb.**

**4**  **Right-click Chapter7.gdb and click New > Feature Class.**

**5**  **In the Name field, type** BusStops.

**6**  **Select Point Features for Type and click Next.**

**7**  **Click NAD_1983_StatePlane_Pennsylvania_South_FIPS_3702_Feet and click Next three times.**

**8**  Type STOP_NAME **as a new field, and select Text as the Data Type.**

The BusStops point features are added to your map document, and you are ready to digitize them.

**9**  **Click Finish and close the Catalog window.**

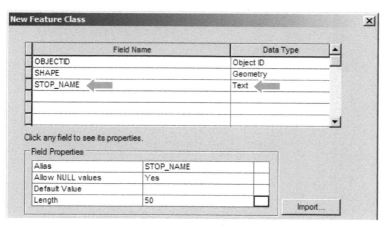

## Turn snapping off

Bus routes in the Middle Hill neighborhood are offset from the street centerlines so you need to turn feature snapping off.

**1**  Click Customize > Toolbars > Snapping.

**2**  Deselect the endpoint and vertex buttons to turn all snapping options off.

## Prepare the map and digitize points

**1**  In the table of contents, click the legend symbol for the BusStops layer. Change the symbol to Circle 2, the color to Mars Red, and the size to 8, and click OK.

**2**  Zoom to the Centre Avenue Route bookmark.

**3**  From the Editor toolbar, click Editor > Start Editing.

**4**  Click BusStops > OK.

**5**  On the Editor toolbar, click the Create Features button.

**6**  Click BusStops in the Create Features panel and Point in the Construction Tools panel.

**7** Click to place the following approximate bus stop points along the Centre Avenue route:

**8** When finished, click Editor > Stop Editing and click Yes to save the BusStops points.

**YOUR TURN**

Digitize bus stops for Bedford/Erin, Webster, Wyle, and Kirkpatrick Streets. Use spatial bookmarks to zoom to the route locations. Enter the fictitious bus stop names (for example, Centre and Soho) for a few random stops. When finished, stop editing and save the edits to your bus stops. Add the CommercialZones polygons that you created in tutorial 7-1. Does it appear that the bus routes and stops serve the commercial zones? Save the map document.

## TUTORIAL 7-4
# Using advanced editing tools

There are several advanced editing tools. Here you'll try the Trace, Cut Polygons, Smooth, Generalize, and Rotate tools, all of which affect the shape of digitized polygons.

## Create a new polygon feature class

1    Open Tutorial7-4.mxd from the Maps folder and save as Tutorial7-4.mxd to the Chapter7 folder.

2    Click Windows > Catalog.

3    In the Catalog tree, browse to Chapter7.gdb.

4    Right-click the Chapter7.gdb folder and click New > Feature Class.

5    In the Name field, type CampusStudyArea and select Polygon Features for Type.

6    Click Next and NAD_1983_StatePlane_Pennsylvania_South_FIPS_3702_Feet as the Current coordinate system.

7    Click Next three times, click Finish, and close Catalog. The new CampusStudyArea feature class is added to your table of contents, and you are now ready to digitize a polygon around Carnegie Mellon University's main campus.

# Trace tool

Tracing is a quick way to create new segments that follow the shapes of other features. Tracing is particularly useful when the features you want to follow have curves or complicated shapes, because snapping is more difficult in those cases.

**1**  On the Editor toolbar, click Start Editing > CampusStudyArea > OK.

**2**  Click the Create Features button.

**3**  Click CampusStudyArea from the Create Features panel and Polygon from the Construction panel.

**4**  Close the Create Features panel.

**5**  On the Editor toolbar, click the Trace button ⟁ ▾.

**6**  Click the intersection of Boundary and Forbes and drag your mouse to the right, move the cursor to the Intersection of Forbes and Morewood, follow Morewood, then Forbes again, Margaret Morrison, Tech, Frew, Schenley, and then back to Boundary.

**7**  When you are close to the beginning intersection of Boundary and Forbes, double-click to finish. You can click the Undo button to start over as needed. The finished polygon nicely follows straight and curved segments. Your final polygon should look like the image shown.

# Cut Polygons tool

Campus planners might want the Morewood boundary to be a separate polygon. The Cut Polygons tool creates two polygons from one original polygon.

**1**   On the Editor Toolbar, click the Cut Polygons tool ⊡.

**2**   Click the left intersection of Forbes and Morewood Streets, and then double-click the right intersection of Forbes and Morewood Streets as shown in the image. The result is two new polygons.

**3**   On the Editor toolbar click Editor > Save Edits.

# Smooth tool

The Smooth tool smooths sharp angles in polygon outlines to improve aesthetic quality.

**1**  Click Bookmarks > Schenley Park.

**2**  On the Editor toolbar, click the Create Features button > CampusStudyArea > Polygon, and click the Straight Segment button.

**3**  Digitize a new polygon with 20 to 25 vertices as shown in the image.

**4**  On the Editor toolbar, click the Edit button and click inside the polygon.

**5**  On the Editor toolbar, click Editor > More Editing Tools > Advanced Editing. This adds a new Advanced Editing toolbar.

**6** On the Advanced Editing toolbar, click the Smooth tool 🔍, type a maximum allowable offset of 10, and click OK. Smoothing adds many shape vertices to create smooth curves between the polygon's vertices.

**7** Click the Edit tool, double-click inside the polygon, and move vertex points to better match the park outline. Add or delete vertex points as needed.

**8** Click outside the polygon to confirm the new shape.

## Generalize tool

Generalizing creates features for use at small scales with less detail while preserving basic shapes.

**1** On the Editor toolbar, click the Edit tool and double-click inside the new traced polygon.

**2** On the Advanced Editing toolbar, click the Generalize tool, 🔲 type a Maximum allowable offset of 50, and click OK. The result is a polygon with fewer vertices, no two of which have a line segment between them of less than 50 feet.

You can click the Undo button to try a different offset.

**3** Click the Edit tool, double-click inside the polygon, and press Delete.

**4** On the Editor toolbar, click Editor > Stop Editing and save your edits.

## Specify a segment angle and length

You have learned how to use a basemap or an aerial image as a guide to digitize new features. You also can digitize features by specifying an exact length and angle. In this exercise, you'll create a performance stage in the middle of Schenley Park that is exactly 80 by 40 feet.

**1** On the Editor toolbar, click Editor > Start Editing > CampusStudyArea > OK.

**2** Click CampusStudyArea in the Create Features panel and Rectangle in the Construction Tools panel.

**3** Click anywhere in the middle of Schenley Park.

**4** Move your cursor to start drawing a polygon, right-click, and click Direction.

**5**   Type 0 and press Enter.

**6**   Right-click and click Length, type 80, and press Enter.

**7**   Right-click and click Width, type 40, and press Enter. Your rectangle is 80 feet long and 40 feet wide and its direction is to the right (ArcMap measure angles counterclockwise with zero being east or to the right).

## Rotate and move a feature

The stage you created is almost finished. Next, you'll rotate and move it to the actual location in the middle of the park.

**1**   Click the Edit tool and click the performance stage rectangle polygon.

**2**   On the Editor toolbar, click the Rotate tool.

**3**   Click on the outline of the rectangle and drag to rotate it to match the location of the performance stage on the aerial map.

**4**   Click the Edit tool and move the stage on top of the stage in the aerial map.

**5**   Click Editor > Stop Editing and click Yes to save your edits.

**6**   Save your map document.

## TUTORIAL 7-5
# Spatially adjusting features

The ArcMap spatial adjustment tools transform, rubber sheet, and edgematch features in a shapefile or geodatabase feature class. In this exercise, you'll transform an outline of a building so that it correctly overlays the building in the aerial image.

## Prepare the map

**1**   **Open Tutorial7-5.mxd from the Maps folder and save as** Tutorial7-5.mxd **to the Chapter7 folder.**

**2**   **Click the Zoom to Extent button to see both the aerial map and the building.** The Hamburg Hall layer, originally a CAD drawing exported as a shapefile, is not in proper alignment or scale with the buildings shown on the aerial image. Next, you'll adjust the building layer so that it properly aligns with the aerial image.

**3**   **In the table of contents, right-click Hamburg Hall > Data > Export Data and save this as a new feature class named** HamburgHall **in Chapter7.gdb.**

**4** Add the new polygon feature to the map document and remove the original Hamburg Hall layer.

## Move the building

**1** On the Editor toolbar, click Editor > Start Editing. Make sure the Create Features layer is Hamburg Hall.

**2** Click the Edit tool, click inside the Hamburg Hall feature, and drag it to the location shown on the image.

**3** Zoom in to better see the Hamburg Hall feature. The building is too large and is upside down.

**YOUR TURN**   Rotate the Hamburg Hall feature 180 degrees.

# Add displacement links

To align the feature with the aerial image, you'll use a transformation tool.

**1** On the Editor toolbar, click Editor > More Editing Tools > Spatial Adjustment. This opens the Spatial Adjustment toolbar.

**2** On the Spatial Adjustment toolbar, click Spatial Adjustment > Adjustment Methods > Transformation - Similarity.

**3** Click the New Displacement Link tool ⚲.

**4** Click the upper-left corner of the Hamburg Hall building feature.

**5** Click the corresponding location on the aerial image.

**6** Continue adding displacement links to the building feature and the aerial image, as shown in the image.

# Edit displacement links

If you select the wrong position on the building or map, you can use the edit displacement tools to adjust your picks.

**1** From the Spatial Adjustments toolbar, click the Select Elements button.

**2** Click one of the displacement links.

**3** Click the Modify Link tool ⚲.

**4** Drag the link to a new position.

**5** Drag the link back to its original location.

> ## YOUR TURN
> Zoom to Hamburg Hall in the aerial image and use the Modify Links tool to more precisely move the displacement links to the corners of the building as shown in the figure.

## Adjust the building

**1** From the Spatial Adjustment toolbar, click Spatial Adjustment > Adjust.

**2** Stop editing and save your edits. ArcMap scales down the Hamburg Hall feature to match the geometry of the feature in the aerial image. If the resulting match is not as good as you would like, select the Hamburg Hall feature, redefine new displacement links, and run the Adjust command again.

**3**   Save your map document and exit ArcMap.

ASSIGNMENT 7-1

# Digitize police beats

Community-oriented police officers are responsible for preventing crime and solving underlying community problems related to crime. Among other activities, these officers walk "beats," which are small networks of streets in specified areas. Often the beats are designed in cooperation with community leaders who help set policing priorities. Beats change as problems are solved and priorities change. Hence, it is good to have the capability to digitize and modify police beats.

In this assignment, you'll digitize two new polyline police beats for the City of Pittsburgh Zone 2 Police District based on street centerlines that make up these beats.

## Get set up

- Rename the folder **\EsriPress\GIST1\MyAssignments\Chapter7\Assignment7-1YourName\** to your name or student ID. Store all files that you produce for this assignment in this folder.
- Create a new map document named **Assignment7-1YourName.mxd** with relative paths.
- Create a new file geodatabase named **Assignment7-1YourName.gdb**.

## Build the map

Import or copy the following data into your file geodatabase:

- \EsriPress\GIST1\Data\Pittsburgh\Zone2.gdb\Streets — line features for Zone 2 Police District streets.
- \EsriPress\GIST1\Data\Pittsburgh\Zone2.gdb\Outline — polygon features for an outline boundary of the Zone 2 Police District. Be sure to export Walkways.dwg into your file geodatabase before attempting to move and rotate these features. Move and rotate will not work using the original CAD drawing.

## Create and digitize new features

In your file geodatabase, create a new polyline feature class named **PoliceBeats** and assign it the same coordinate system as the Zone2 features, NAD_1983_StatePlane_Pennsylvania_South_FIPS_3702_Feet. Create a new text field named **BeatNumber**. Digitize line segments for the police beats using the guidelines that follow for what streets make up the beats. Populate the fields as Beat1 and Beat2 for each line segment making up that beat.

*Hints:* Open the feature attribute table for the streets. Move the table so you can see both the table and the streets on the map. Sort the table by field, and NAME, and make multiple selections for a given beat in the table by simultaneously holding down the CTRL key and clicking rows corresponding to the beat's street segments. The streets layer is a TIGER file map with TIGER-style address number data, so look for street number ranges in the following fields: L_F_ADD, L_T_ADD, R_F_ADD, and R_T_ADD. With all streets for a beat selected, digitize lines for streets making up beats in the new line features for Beat1 and Beat2. Use the snapping endpoint and vertex tools using street centerlines as a guide.

**Street centerline guides for Beat #1 (21 streets)**

- 1 through 199—17th St (four segments)
- 1 through 99—18th St (two segments)
- 1 through 199—19th St (one segment)
- 1 through 199—20th St (four segments)
- 1 through 99—Colville St (one segment)
- 1700 through 1999—Liberty Ave (one segment)
- 1700 through 1999—Penn Ave (three segments)
- 1700 through 1999—Smallman St (four segments)
- 1700 through 1999—Spring Way (one segment)

**Street centerline guides for Beat #2 (20 streets)**

- 100 through 299—7th St (two segments)
- 1 through 299—8th St (three segments)
- 100 through 299—9th St (four segments)
- 800 through 899—Exchange Way (one segment)
- 700 through 899—Ft Duquesne Blvd (three segments)
- 700 through 899—Liberty Ave (three segments)
- 100 through 199—Maddock Pl (one segment)
- 700 through 899—Penn Ave (three segments)

Be sure to double-click to create individual line segments as indicated.

## Requirements

- Create a layout with three data frames, renaming the data frames and layers. Using Unique Values, show the beats with line widths 1.5 and bright, distinctive colors (Mars Red and Ultra Blue), and show streets as lighter ground features.
- In data frame 1, show an overview map zoomed to the Police Zone 2 outline, with existing streets and the newly digitized beats labeled as **Beat1** and **Beat2** using a callout label from the Draw toolbar.
- In data frames 2 and 3, include maps zoomed to Beat1 (data frame 2) and Beat2 (data frame 3) at the same scale. Label the streets in the detailed maps.
- Include scale bars in miles for the overview map and in feet for the individual beats' maps. Include your name as the author and the current date.
- Click File > Export Map and save your map layout as **Assignment7-1YourName.jpg** with 150 dpi resolution.

# Use GIS to track campus information

GIS is a good tool to create campus information or emergency preparedness maps. These maps can be used in many organizations that have large campuses or complicated buildings. For example, airports, hospitals, office parks, or colleges and universities can be confusing, especially to new students and visitors. Visitors might also be unaware of the locations of features that would be useful for emergency purposes.

In this assignment, you'll create spatially modified and new features for sidewalks, parking lots, handicap parking, and emergency management service locations. These features are shown in the image for this assignment, and you can visit Carnegie Mellon University's visitor map for additional parking and emergency service information: https://www.cmu.edu/visit/maps-parking-transportation.html.

## Get set up

- Rename the folder **\EsriPress\GIST1\MyAssignments\Chapter7\Assignment7-2YourName\** to your name or student ID. Store all files that you produce for this assignment in this folder.
- Create a new map document named **Assignment7-2YourName.mxd** with relative paths.
- Create a new file geodatabase named **Assignment7-2YourName.gdb**.

## Build the map

Import or copy the following data and raster image into your file geodatabase:

- \EsriPress\GIST1\Data\Pittsburgh\City.gdb\PghStreets—street centerline features for Pittsburgh.
- \EsriPress\GIST1\Data\CMUCampus\Walkways.dwg—CAD drawing of campus sidewalks (originally extracted from CMU's Facilities Management Services campus drawing). Use the line features of this drawing file.
- \EsriPress\GIST1\Data\CMUCampus\CMUCampus.tif—USGS ortho image of Carnegie Mellon University's campus.

## Create new features

In your file geodatabase, create the following feature classes using the same map projection as PghStreets features, NAD_1983_StatePlane_Pennsylvania_South_FIPS_3702_Feet:

- **ParkingLots** (polygons) with a new text field named **LOT_NAME**
- **HandicapParking** (points)

• **EMS_Sites** (points) with a new text field named **NAME**

## Requirements

• Use the Move and Rotate Vertex tools to display the CMU sidewalks to the approximate locations shown as white lines in the next figure's aerial image.
• Digitize the five parking lots shown in the image and enter the corresponding parking lot names. Label the lot name on your map. Display the parking lots as semitransparent polygons so you can see the parked cars in the aerial image. Be sure to double-click to create individual line segments as indicated.
• Digitize 11 handicap parking locations as shown in the image using the Esri "Handicapped2" symbol.
• Digitize EMS locations for Health Services and University Police as shown and enter corresponding names in the attribute table. Display the points as unique symbols using the Esri "Cross 4" and "Asterisk 2" symbols.
• Display PghStreets with no color but labeled using a small halo mask.
• Create a map layout zoomed to the above features at a scale of 1:4,000. Include a title, legend, map scale using feet, your name, and the current date.
• Click File > Export Map and save your map layout as **Assignment7-2YourName.jpg** with 150 dpi resolution.

# 8

# Geocoding

Geocoding is the process used to plot address data as points on a map. You can geocode addresses to different levels such as ZIP Codes or streets, depending on the type of address data you have or wish to map. In this chapter, you'll learn to geocode using source tables of address data and location reference data obtained from the US Census Bureau. You'll also learn how to fix errors in both the source and reference data you use for geocoding.

Learning objectives

- Geocode by ZIP Code
- Geocode by street address
- Correct address data interactively
- Edit street map layers to improve geocoding
- Use an alias table for place-names

## TUTORIAL 8-1
## Geocoding data by ZIP Code

Geocoding to ZIP Codes is a common practice for many organizations because ZIP Code data is often available in client and other databases. Furthermore, for marketing and planning, it is often sufficient to study the spatial distribution of clients by ZIP Code. ZIP Code areas lack an underlying design principle except for delivery of mail, so interpretation of results is sometimes limited. In these tutorials, you'll geocode attendees for an art event sponsored by an arts organization in Pittsburgh, Pennsylvania, named FLUX. The event planners of FLUX would like to know where function attendees reside for planning and marketing activities.

# Open and examine the starting map document

**1** **Start ArcMap and open Tutorial8-1.mxd from the Maps folder.** The map document includes the two needed inputs for geocoding, the ZIP Code map of Pennsylvania, and the data table Attendees.

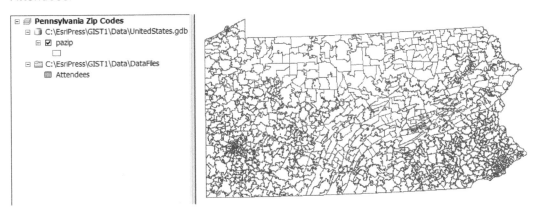

**2** **Save the map document to the Chapter8 folder of MyExercises. Click the List By Source button at the top of the table of contents.**

**3** **In the table of contents, right-click the Attendees table and click Open.** The table contains the addresses, including 5-digit ZIP Code, and ages of all attendees of two recent FLUX events. Notice that two of the first six attendees are out of state and thus will not geocode with the Pennsylvania ZIP Code map. As a supplement, before closing the table, let's tabulate how many such records there are. That informs the performance assessment of geocoding within Pennsylvania.

| Attendees | | | | | | |
|---|---|---|---|---|---|---|
| OID | Date_ | Address | City | State | ZIP_Code | Age |
| 0 | 20090629 | 2415 1ST AVE | SACRAMENTO | CA | 95818 | 27 |
| 1 | 20090629 | 224 NORTH ST | STERBENVILLE | OH | 43952 | 32 |
| 2 | 20090629 | PO BOX 622 535 4TH ST | MARIANNA | PA | 15345 | 32 |
| 3 | 20090629 | 5126 JANIE DRIVE | PITTSBURGH | PA | 15227 | 34 |
| 4 | 20090629 | 305 AVENUE A | PITTSBURGH | PA | 15221 | 40 |
| 5 | 20090629 | 1431 CRESSON ST | PITTSBURGH | PA | 15221 | 26 |

**4** **Using the Table Options button, click the Select By Attributes button, double-click "State" in the top panel, click the = button, click Get Unique Values, scroll down and double-click 'PA,' and click Apply.** Information at the bottom of the table indicates that 1,124 out of 1,265 records (89 percent) are for Pennsylvania, so that is the maximum number of points that could be geocoded. If there are missing or incorrect ZIP Codes for Pennsylvania, the number geocoded will be less.

**5** **Using the Table Options button, click the Clear Selection button and close the table.**

# Create an address locator for ZIP Codes

The geocoding process requires several settings and parameters. Rather than have you specify them interactively each time you geocode data, ArcMap has you save settings in a reusable file, called an *address locator*. Included in the settings is a pointer to the reference data you use to geocode the attendee data, the PAZip (Pennsylvania ZIP Code) layer that is currently in your map document.

**1**   In the Catalog window, navigate to the Chapter8 folder of MyExercises, right-click the Chapter8 folder, and click New and Address Locator.

**2**   In the Create Address Locator window, click the browse button for the Address Locator Style, click US Address–ZIP 5-Digit, click OK, and ignore the warning icon and message.

**3**   Click the list arrow for Reference Data, and click pazip.

**4**   For the Output Address Locator, change the file name from PAZip_CreateAddressLocator to PAZipCodes.

**5**   Click OK and wait until Catalog informs you that the address locator is created. PAZipCodes appears as an address locator under Chapter8.gdb in Catalog.

**6**   Hide the Catalog window.

# Geocode records by ZIP Code

**1**   In ArcMap, click Customize > Toolbars > Geocoding. The Geocoding toolbar appears.

**2**   On the Geocoding toolbar, click the Geocode Addresses button 🖳, click PAZipCodes to select it, and click OK. Click the option button for Single Field, select ZIP_Code for the Complete ZIP Code field, and change the output name to \EsriPress\GIST1\MyExercises\ Chapter8\Chapter8.gdb\AttendeesZIP.

**3**  **Click OK.** ArcMap geocodes the addresses by ZIP Code with 86 percent of the records mapped — less than the 89 percent of addresses that are in Pennsylvania, as expected.

**4**  **Click Close.** ArcMap adds the geocoding results to the map with point markers at the centroids of ZIP Codes that have one or more attendees. As you might expect, attendees cluster around southwestern Pennsylvania near the location of the FLUX events, but attendees come from all over the state.

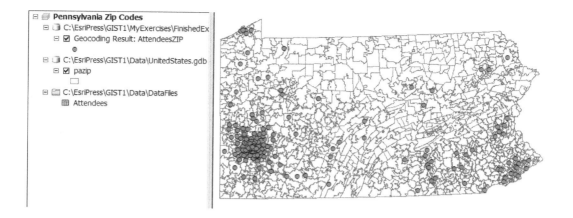

## Count geocoded records by ZIP Code

You can get a better understanding of the geocoded output by next taking an extra step, aggregating geocoded points to obtain a count of attendees per ZIP Code area.

**1** In the table of contents, right-click GeocodingResult: AttendeesZIP and click Open Attribute Table.

**2** Right-click the Match_Addr column header, click Summarize, change the output table to \ EsriPress\GIST1\MyExercises\Chapter8\Chapter8.gdb\CountAttendees, **and click OK > Yes.**

**3** Close the geocoding results table.

**4** In the table of contents, right-click CountAttendees and click Open.

**5** Right-click Count_Match_Addr column header and click Sort Descending. In total, there are 1,487 ZIP Codes in the state (which you can determine by opening the attribute table of PAZip), but only 22 ZIP Codes with 10 or more attendees (1.5 percent) and 90 (6.1 percent) that have 2 or more attendees. This is valuable information for FLUX for targeting marketing activities.

**6** Close the table.

## Fix and rematch ZIP Codes

**1** In the table of contents, click Geocoding Result: AttendeesZIP and click the Review/ Rematch Addresses button 🔌 on the Geocoding toolbar. The Interactive Rematch window shows each unmatched record individually and allows you to manually edit the address values.

**2** For Show results, select Unmatched Addresses, scroll over to the Address field, and scroll down to find the record whose address value is 414 South Craig Street, and select that record.

**3**  Scroll horizontally (and adjust field widths by dragging their header boundaries) so that you can see Address, City, State, and ZIP_Code fields. Notice that ZIP_Code is missing for this record. That address's ZIP Code is 15213.

**4**  With the 414 SOUTH CRAIG STREET row selected, type 15213 in the Complete ZIP Code field (top of lower-left panel) and press Tab on your keyboard. The Candidate panel shows one candidate with a perfect score of 100.

**5**  Click the Match button. The count of matched addresses goes up by one, from 1,090 to 1,091, as ArcMap is successful with that point. With some research, it is usually possible to make similar additions.

> ## YOUR TURN
> Use the US Postal Service's ZIP Code lookup website, https://tools.usps.com/go/ZipLookupAction!input.action, to find the ZIP Code for the record for 11244 Azalea Dr, Pittsburgh, PA (it's 15235). Then use Interactive Rematch to match the address. Close the Interactive Rematch window, and save your map document when finished.

## TUTORIAL 8-2
# Geocoding data by street address

In this exercise, you'll again geocode the FLUX attendee records, but this time at the street level for Pittsburgh. In this case, you'll incorporate ZIP Code into the address locator, because some addresses may have the same house number and street name but be in different ZIP Code areas. This happens frequently in study areas that have two or more municipalities, such as in a county. In the FLUX case, there are address records from other cities and states, so ZIP Code plays an important role, considering the preponderance of Pittsburgh records, to eliminate matches with non-Pittsburgh street addresses that have the same street address as in Pittsburgh.

## Examine address data and a street map

**1**  Open Tutorial8-2.mxd from the Maps folder and save the map document in the MyExercises Chapter8 folder.

**2** Click the List By Source button at the top of the table of contents. Open the attribute tables for PghStreets and Attendees and review their contents, especially addresses. You will find that PghStreets has TIGER-style street address data, with starting and ending house numbers for each street segment. The table, Attendees, has street address in one field, Address, plus City, State, and ZIP_Code in their own fields. Only Address and ZIP_Code are necessary for geocoding, because with ZIP_Code you can look up city and state.

**3** Close the tables after you are finished reviewing them.

## Create an address locator for streets with a zone

**1** Click the Catalog icon in the right edge of the ArcMap window (or, if the icon is not there, click Windows > Catalog).

**2** Expand the Chapter8 folder of MyExercises and right-click New > Address Locator.

**3** In the Create Address Locator window, click the browse button for the AddressLocator Style, click US Address—Dual Ranges, click OK, and ignore the warning icon and message.

**4** Select PghStreets for Reference Data, and change the Output Address Locator name to PghStreets. Notice that Catalog identifies all fields that it needs for the geocoding process in the lower panel of the Create Address Locator window. If it was unsuccessful in doing so, you would have to click in a cell on the right and select the needed field name from the table. Field names that start with an asterisk are required.

**5** Click OK and wait until Catalog informs you that the address locator is created.

**6** Expand Chapter8 in the Catalog tree and double-click PghStreets. Expand all items in the resulting property sheet (see figure). In the Place name alias table item, you can associate an alias table with the locator. It would contain place-names such as PNC Ballpark and associated street addresses such as 115 Federal Street. With an alias table, ArcMap makes a pass through address data, replacing place-names with their street addresses. In Outputs, you can have x and y coordinates written to the output attribute table. In Geocoding options are parameters with default values that determine how strict or lax addresses must compare in the source and reference datasets to be considered a match. Work Assignment 8-3 to learn how to modify these parameters. Also in this element, you see the allowable connectors for street intersection addresses, such as Oak St & Pine Ave, are currently the "&", "@", and "|" characters, and the words "and" and "at." If your data has a different separator, you could type it here. Also, the point that ArcMap assigns to an address has a 20-foot side offset on the correct side of the street (left or right). You can change the offset to another value if desired. Offsets are desirable, especially for aggregating geocoded data points up to counts by area, such as neighborhoods. Areas tend to use street centerlines as boundary lines, so an offset ensures that ArcMap will count each point once, in the correct polygon. If on the centerline, a point gets double-counted for polygons that share the street segment as a boundary line.

**7** Click Cancel to close the Address Locator Properties window, and hide the Catalog window.

## Interactively locate addresses

Before geocoding the Attendees table to the streets layer, you'll try out your locator with ArcMap's Find tool to locate individual addresses. The Find tool that you'll use has the same methodology as geocoding to transform street address data into a point on the map. It matches the address you type in with similar data stored as attribute data in the street centerline map. It does the matching by finding good candidates and then computing a match score for each. Then for each identified problem or flaw of a candidate in matching the desired address, ArcMap subtracts a penalty from the match score. The candidate with the perfect score, 100, or highest score above a threshold value is chosen as the geocoded point.

**1** On the Tools toolbar, click the Find button 🔍.

**2**  **Click the Locations tab.** ArcGIS should automatically find your PghStreets locator. If not, click the browse button for locator, and browse to and select your PghStreets locator in Chapter8. gdb.

**3**  **Type** 3609 Penn Ave 15201 **in the Full Address field and click Find.** The locator finds the address, briefly flashing it on the map. The address locator works well.

**4**  **Right-click the matched address of the lower panel to open a context menu and click the Add Labeled Point option.** A point appears at the corresponding point on the map, along with a label for the street address.

**5**  **Close the Find window.**

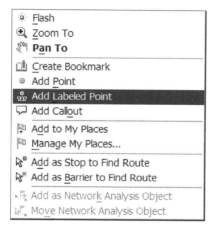

**YOUR TURN**

Use the Find tool to locate the following addresses. Add the best match in each case to the map as a labeled point.

    1920 S 18th ST 15203
    255 Atwood ST 15213
    3527 Beechwood BLVD 15217

The finished map has some editing of the labels, which is optional for you. You can ungroup a label and its point by clicking the label to select the graphic, right-clicking the graphic, and clicking Ungroup. Then you can format the font and point color by right-clicking an element, clicking Properties, and clicking Change Symbol.

When finished, delete the label and point graphics by dragging a rectangle around them to select them and press Delete on your keyboard.

## Geocode address data to streets

We do not expect ArcMap to geocode a high percentage of matched records, because the streets reference layer is only for Pittsburgh while many attendees live outside of Pittsburgh. Also, note that your matched addresses and statistics may not match those shown here because the Esri software developers frequently improve ArcGIS locators, even with releases of software service packs.

**1** If the Geocoding toolbar is not already open, click Customize > Toolbars > Geocoding. Note: If you get the error message "Geocoding an address table with a definition query or selection is not supported," open and close the Attendees table.

**2** On the Geocoding toolbar, click the Geocode Addresses button, select PghStreets, and click OK.

**3** Type or make selections as shown in the image, with Attendees for Address table, ZIP_ Code under Address Input Fields, and AttendeeStreets for Output shapefile or feature class. Do not click OK.

**4** **Click Geocoding Options.** This is a user dialog box controlling the behavior of address matching and its outputs. You saw several of these options earlier. Here you can choose to add the X and Y coordinates of matched points to the output map layer's attribute table; however, leave all default settings as found here.

**5** **Click Cancel and click OK.** ArcMap only matches about 550 (43 percent) of records, a relatively low number as expected, because there are many addresses outside of Pittsburgh as well as unmatched addresses in Pittsburgh. You can count 794 Pittsburgh records in the Attendees table, so the match rate for Pittsburgh is low, 550 out of 794, or 70 percent. Sometimes street address data collected in surveys is low quality as seems to be in the case here.

**6**   **Click Close and save your map.** The residences of attendees have distinctive spatial patterns across Pittsburgh neighborhoods, which could inform marketing campaigns for increasing attendance. It looks like most attendees come from only 15 to 20 of the 91 neighborhoods in Pittsburgh.

## TUTORIAL 8-3

# Correcting source addresses using interactive rematch

Although ArcMap did not match many of the addresses in the Attendees table to the PghStreets layer because they are outside Pittsburgh, some addresses in Pittsburgh did not match because of spelling errors or data omissions in the input table. Making corrections to the address data depends on the user's knowledge of local streets and addresses. In this tutorial, you'll use the ArcMap interactive review process to correct and then match a few of the unmatched records.

## Rematch interactively by correcting input addresses

**1**   **Save your map document as** Tutorial8-3.mxd **to the Chapter8 folder.**

CHAPTER 8: GEOCODING    275

**2** Click Geocoding Result: AttendeesStreets in the table of contents. On the Geocoding toolbar, click the Review/Rematch Addresses button. You can increase the height of the panel with addresses by finding and dragging horizontal boundary lines.

**3** In the Show results field, select Unmatched Addresses.

**4** Horizontally scroll across the fields so you can see the Address, City, State, and ZIP_Code fields, and then make those fields narrower by dragging vertical boundaries between fields so that you can see all their values.

**5** Right-click the Address header in the data panel and click Sort Ascending.

**6** Scroll down to the record with address 120 W 9th STREET and select that record. ArcMap did not match this address because its direction is wrong and its ZIP Code is wrong. The correct direction is S, not W and the ZIP Code should be 15203.

**7** Change the direction from W to S in the Street or Intersection field and type 15203 in the ZIP Code field of the lower-left panel, press the Tab key, and click Match. ArcMap matches the records and the count of Matched addresses advances by one.

---

## YOUR TURN

Rematch an additional record, 1715 CENTER AVE. The problem here is that CENTER is spelled wrong (it should be CENTRE) and the ZIP Code is wrong (it should be 15219). Make the corrections in the lower-left panel of Interactive Rematch and click Match. The number matched will advance. Leave the Interactive window open.

---

## Rematch interactively by pointing on the map

Sometimes you will have unmatched records that you can find on the map using external information or expert knowledge, but your reference data (street map) simply will not have a corresponding address or street. Often, too, TIGER street maps have missing house numbers or other data for street segments so that address matching is not possible. In such cases, ArcMap lets you point on the map to geocode.

**1** If the Interactive Rematch window is not open, click Geocoding Result: AttendeesStreets in the table of contents. Then on the Geocoding toolbar, click the Review/Rematch Addresses button. Select Unmatched Addresses for the Show results field.

**2** Sort the address data by ObjectID, scroll down to the record with ObjectID 1010, and click that record to select it. Although this record has only a ZIP Code value, 15221, for address data, suppose that a comment field of the original survey data mentioned that the attendee lived on Canada Way. That street is only one block long in Pittsburgh, so you decide to point to the middle of that street to address match the point.

**3** On the Tools toolbar, click the Zoom In tool and zoom in to the eastern portion of Pittsburgh as seen in the image. Highlighted is Canada Way. If you have difficulty locating that street, you can use any street segment in its vicinity for practice here.

**4** On the bottom of the Interactive Rematch window, click Pick Address from Map, move your cursor to the middle of Canada Way, right-click, click Pick Address, and scroll to the left in the address panel of the Interactive Match window. ArcMap adds the address as a point on the map where you clicked. It sets the record's Status to *M* for matched and sets the Match_ type to PP for picked point (for other records, *A* is the code for address matched).

**5** Close the Interactive Rematch window and click the Full Extent button on the Tools toolbar. Save your map.

## TUTORIAL 8-4
# Correcting street reference layer addresses

In this tutorial, you'll learn how to find and fix an incorrect address in a reference street layer used for geocoding. To do this, you'll examine unmatched user addresses, identify candidate streets for revisions, and examine the attributes of the streets to look for misspellings or data omissions.

# Open a map document

**1** **Open Tutorial8-4.mxd from the Maps folder.** Tutorial8-4 contains a table of clients and a street centerlines layer for Pittsburgh's Central Business District.

**2** Save the map document to the Chapter8 folder.

**3** **Right-click Clients in the table of contents and click Open.** The data table has addresses for all 27 records, including ZIP Code. Notice that the last four records have place-names instead of street addresses. Later in the tutorial you'll use an alias table to geocode those locations. The alias table includes street addresses for place-names.

| OBJECTID * | LAST_NAME | FIRST_NAME | ADDRESS | ZIP |
|---|---|---|---|---|
| 1 | Roberts | Louisa | 133 Seventh St | 15222 |
| 2 | Johnson | Tammy | 615 Penn Ave | 15222 |
| 3 | Clark | Robert | 309 Ross St | 15222 |
| 4 | Peterson | Jennifer | 118 6th Street | 15222 |
| 5 | Young | Mike | 490 Penn Ave | 15222 |
| 6 | Thompson | Samantha | 711 Liberty Ave | 15222 |
| 7 | Reed | Rhonda | 111 Hawksworth | 15222 |
| 8 | Baker | Sally | 900 Smallman St | 15222 |
| 9 | Wilson | Laura | 599 Smithfield St | 15222 |
| 10 | Jenkins | Amy | 701 Grant Street | 15222 |
| 11 | Kelly | Tabitha | 900 Lib Ave | 15222 |
| 12 | Riley | Jennifer | 777 Illini Drive | 15222 |
| 13 | Smith | Emily | 341 Stanwix St | 15222 |
| 14 | Williams | Polly | 109 Washington Pl | 15222 |
| 15 | Davis | Kelly | 1100 Liberty Ave | 15222 |
| 16 | Dobbins | Joshua | 2 S. Market Place | 15222 |
| 17 | Perry | Catherine | 651 Forbes Ave | 15222 |
| 18 | Nelson | Joseph | 241 Forbes Ave | 15222 |
| 19 | Miller | Matthew | 923 French St | 15222 |
| 20 | Lawson | Todd | 295 Wood St | 15222 |
| 21 | Welch | Karen | 301 5th Ave | 15222 |
| 22 | Sigler | Dan | 119 9th Avenue | 15222 |
| 23 | Hampton | Frances | 401 1st Ave | 15222 |
| 24 | Peters | Amanda | One PPG Place | 15222 |
| 25 | Franklin | John | One PPG Place | 15222 |
| 26 | Burns | Anthony | Two Gateway Center | 15222 |
| 27 | Smith | George | Two PPG Place | 15222 |

**4**   Close the table.

## Create an address locator for CBD streets

**1**   Click the Catalog icon in the right edge of the ArcMap window (or if the icon is not there, click Windows > Catalog).

**2**   Expand the Chapter8 folder of MyExercises, right-click the Chapter8 folder, and click New > Address Locator.

**3**   In the Create Address Locator window, click the browse button for the Address Locator Style > US Address–Dual Ranges > OK, and ignore the warning icon and message.

**4**   Select CBDStreets for the Reference Data.

**5**   Change the locator name to PghCBDStreets in the Output Address Locator field.

**6**   Click OK.

**7**   Hide the Catalog window.

## Geocode clients' addresses to CBD streets

**1**   If the Geocoding toolbar is not open, click Customize > Toolbars > Geocoding.

**2**   On the Geocoding toolbar, click the Geocode Addresses button. Note: If you get the error message, Geocoding an address table with a definition query or selection is not supported, open and close the Clients table.

**3**   Select PghCBDStreets and click OK.

**4**   In the Geocode Addresses window, for Output shapefile or feature class, save the result as CBDClients and click OK. ArcMap matches 12 (44 percent) of the 27 records.

**5** Click Close.

## Identify a problem street segment record using Review/ Rematch Addresses

**1** In the table of contents, click Geocoding Result: CBDClients.

**2** On the Geocoding toolbar, click the Review/Rematch Addresses button. Select Unmatched Addresses for the Show results field.

**3** Scroll to the right in the unmatched addresses, right-click the ADDRESS column header, and click Sort Ascending.

**4** Scroll down and select the record with ADDRESS 490 Penn Ave. Click the Geocoding Options button, type 0 for the Minimum candidate score, and click OK. There are many candidate street matches in the lower panel of the Interactive Rematch window. The closest match is 500 PENN AVE 15222.

**5** Select 500 PENN AVE 15222 in the Candidates panel and zoom in to the area with the yellow point marker on the map. Street address numbers increase from left to right in the CBD, and you can see in the lower right of the Interactive Rematch window that the best candidate's lowest number is 500. So the desired street segment must be to the immediate left of the yellow point marker.

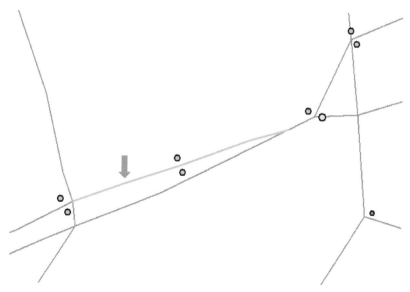

**6** On the Tools toolbar, click the Select button and click the street segment indicated in the figure for step 5.

**7** Close the Interactive Rematch window.

**8**  Open the CBDStreets attribute table and click the Selected button at the bottom of the table.

**9**  Scroll to the right in the table to see that the selected street segment's record is missing its TIGER-style street numbers (from and to, left and right street numbers). Leave the table open.

| RECNUM | L_F_ADD | L_T_ADD | R_F_ADD | R_T_ADD | FCC |
|--------|---------|---------|---------|---------|-----|
| 24924586 | | | | | A35 |

## Edit a street record

Suppose that you have obtained valid numbers for the street segment's missing attributes: 498 to 474 on the left side and 499 to 475 on the right side. You can use the ArcMap Editor toolbar to enter those values.

**1**  On the Menu bar, click Customize > Toolbars > Editor.

**2**  Click the list arrow on the Editor toolbar. Click Start Editing > CBDStreets > OK.

**3**  In the CBDStreets table, type the following values: L_F_ADD=498, L_T_ADD=474, R_F_ADD=499, and R_T_ADD=475.

**4**  Click the list arrow on the Editor and click Save Edits. Click the list arrow again and click Stop Editing.

**5**  Close the CBDStreets table and the Editor toolbar. Now 490 Penn Ave., one of the unmatched client addresses, will geocode the next time you attempt to rematch the addresses.

First, however, you must rebuild the PghCBDStreets Locator so that it includes the edits you just made to the reference CBDStreets layer.

## Rebuild a street locator

If you modify a location reference layer after you built a locator, you must rebuild the locator.

**1**  Unhide or open Catalog.

**2**  If necessary, expand the Chapter8 folder of MyExercises.

**3**  Right-click PghCBDStreets and click Rebuild > OK. If you encounter an error in the rebuild process, close and restart ArcMap.

**4**  Hide Catalog.

## Rematch interactively using an edited street segment

**1** In the table of contents, click Geocoding Result: CBDClients.

**2** On the Geocoding toolbar, click the Review/Rematch Addresses button, select Unmatched Addresses for the Show results field, and select the 490 Penn Ave. address. Now there is a 100 score candidate for geocoding, for exactly the right address, 490 PENN AVE 15222, given your updated street segment.

**3** In the Interactive Rematch window, click the 100 score candidate record > Match. ArcMap matches the record, and the number of matched records increases from 12 to 13. In addition, any time you use CBDStreets in the future for geocoding, ArcMap will successfully geocode any input address records for the edited street segment.

**4** Close the Interactive Rematch window.

**5** Save your map document.

## TUTORIAL 8-5

# Using an alias table

Some places are commonly located by their landmark names instead of their street addresses. For example, the White House may be listed in a table as "White House" instead of 1600 Pennsylvania Avenue NW, Washington, DC 20500. In the following exercise, you'll use an alias table to geocode records that are identified by their landmark name rather than their street address.

## Add an alias table and rematch addresses

**1** Save your map document as Tutorial8-5.mxd.

**2** Click the Add Data button and add the BldgNameAlias table from Data > Pittsburgh > CBD. gdb.

**3** In the table of contents, right-click BldgNameAlias and click Open. The table contains the alias names and street addresses for three records. You can create such tables in Microsoft Excel, in Notepad with comma delimiters, or in other packages, and import them into a file geodatabase using Catalog.

| OBJECTID * | BLDGNAME | ADDRESS | ZIP |
|---|---|---|---|
| 1 | One PPG Place | 200 4th Ave | 15222 |
| 2 | Two PPG Place | 200 4th Ave | 15222 |
| 3 | Two Gateway Center | 197 Stanwix St | 15222 |

**4**  Close the alias table.

**5**  Open Catalog. Right-click the PghCBDStreets locator in the Chapter8 folder and click Properties.

**6**  Expand Place name alias table and select BldgNameAlias for Alias Table and BLDGNAME for Alias Field. Click OK and hide Catalog.

**7**  In the table of contents, click Geocoding Result CBDClients. Click the Review/Rematch Addresses button on the Geocoding toolbar. Select Unmatched Addresses for the Show results field.

**8**  Scroll down in the address records, click the record with ObjectID 21, and scroll to the right to see the address—One PPG Place. Now that you have an alias for that building, there is a candidate with a 100 score.

**9**  Click the candidate with score 100 and click Match. Similarly, match the last three records.

**10**  Close the Interactive Rematch window, save your map, and close ArcMap.

---

## ASSIGNMENT 8-1

# Geocode household hazardous waste participants to ZIP Codes

Many county, city, and local environmental organizations receive inquiries from residents asking how they can dispose of household hazardous waste (HHW) materials that cannot be placed in regular trash or recycling collections. Homeowners continually search for environmentally responsible methods for disposing of common household products such as paint, solvents, automotive fluids, pesticides, insecticides, and cleaning chemicals.

The Pennsylvania Resources Council (PRC) ( http://www.prc.org/) is a nonprofit organization dedicated to protecting the environment. The PRC facilitates meetings, organizes collection events, spearheads fundraising and volunteer efforts, and develops education and outreach materials in response to the HHW problem.

At each disposal event, the PRC collects residence data from participants. In this assignment, you'll geocode participants by ZIP Code for a recent Allegheny County event with collection of hazardous waste that took place in the southeastern portion of Allegheny County's North Park. The beauty of ZIP Code data is that everyone knows their five-digit ZIP Code so that you can get highly accurate and nearly complete data that geocodes well.

# Get set up

- Rename the folder **\EsriPress\GIST1\MyAssignments\Chapter8\Assignment8-1YourName\** to your name or student ID. Store all files that you produce for this assignment in this folder.
- Create a new map document named **Assignment8-1YourName.mxd** with relative paths. Use the UTM NAD1983 Zone 17N projection.
- Create a new file geodatabase named **Assignment8-1YourName.gdb**. Save all new files that you create in this file geodatabase.

# Build the map

Add the following spatial data to your map document:

- \EsriPress\GIST1\Data\UnitedStates.gdb\HHWZIPCodes— table of five-digit ZIP Codes for an HHW Allegheny County event provided by the PRC. Note that the PRC suppressed all attributes except ZIP Code to protect confidentiality.
- \EsriPress\GIST1\Data\UnitedStates.gdb\PAZIP— polygon features of Pennsylvania ZIP Codes used for address matching.
- \EsriPress\GIST1\Data\UnitedStates.gdb\PACounties — polygon features of Pennsylvania counties.
- \EsriPress\GIST1\Data\AlleghenyCounty.gdb\Parks—polygon features of Allegheny County parks.

Geocode HHW ZIP Codes:

- Create an address locator to use for geocoding HHW participants to ZIP Codes, **HHWZIPLocator**.
- Geocode a new map layer, **HHWZIPCodeResidences**.
- Spatially join HHWZIPCodeResidences to PAZIP. *Hint:* Right-click PAZIP, click Joins And Relates > Join and select join data from another layer based on spatial location.

# Requirements

- Symbolize your map (including a choropleth map) in a layout showing the number of Household Hazardous Waste participants by ZIP Code in Pennsylvania. Use five quantiles for the choropleth map. Add the PA County polygons as a thick dark outline. Label counties with county names. Add a second copy of PAZIP with outline and white fill to complete ZIP Codes throughout Pennsylvania where there were no HHW client residences. *Hint:* Select North Park and place Parks below the choropleth map, to see only North Park's location. Add text to your layout to label North Park.
- Export a layout with title, map, and legend to a file named **Assignment8-1YourName.jpg**. Include the layout image in a Word document, **Assignment8-1YourName.docx**, in which you describe residence patterns of HHW event attendees relative to North Park. Include a statement about the top 20 percent of clients. Roughly speaking, how far were they willing to travel; 5, 10, 15, or 20 miles? Include geocoding match statistics in your Word document.

## ASSIGNMENT 8-2

# Geocode immigrant-run businesses to Pittsburgh streets

As the 2010 Census shows, immigrants are becoming one of the most salient indicators of growth and wealth in a region. By looking at the immigrants who live in a city and analyzing where they decide to set up their businesses, city planners can investigate why certain neighborhoods are more immigrant-friendly than others, and turn their focus on the discovered qualities that make a neighborhood open and diverse. In this assignment, you'll geocode and study immigrant-run high-tech firms, restaurants, and grocery stores in Pittsburgh.

## Get set up

- Rename the folder **\EsriPress\GIST1\MyAssignments\Chapter8\Assignment8-2YourName\** to your name or student ID. Store all files that you produce for this assignment in this folder.
- Create a new map document named **Assignment8-2YourName.mxd** with relative paths. Use a state plane projection.
- Create a new file geodatabase named **Assignment8-2YourName.gdb**. Save all new files that you create in this file geodatabase.

## Build the map

Add the following spatial data to your map document:

- \EsriPress\GIST1\Data\Pittsburgh\City.gdb\ImmigrantBusinesses— sample of immigrant-run businesses in Pittsburgh. Attributes include street address, city, state, ZIP Code, and type of business (Firm, Grocery, and Restaurant).
- \EsriPress\GIST1\Data\Pittsburgh\City.gdb\PghStreets — line features of Pittsburgh street centerlines.
- \EsriPress\GIST1\Data\Pittsburgh\City.gdb\Neighborhoods — polygon features of Pittsburgh neighborhoods.

## Geocode immigrant businesses

- Create an address locator, named **BusinessStreetsLocator**, to use when geocoding immigrant-run businesses to streets.
- Use US Address—Dual Ranges as the type of locator and PghStreets as the reference data. As address input fields, use both ADDRESS and ZIP.
- Save the geocoded businesses as **ImmigrantBusinesses**.

## Rematch unmatched ZIP Codes

You should get about 30 percent that do not match. Reasons include wrong ZIP Code data, place-names instead of street addresses, and incorrect or misspelled street addresses.

- Open the Interactive Rematch window and show results for Unmatched Addresses. The address, 700 Technology Dr, 15230, should be the first unmatched address listed. Start with it and work your way down the list until you get five more good matches.
- Use Internet sites such as https://www.usps.com/ or https://maps.google.com/ to rematch five unmatched addresses.
- When you find potential corrections, try making changes in the lower-left panel of the Interactive Rematch window. If you get a good match, click the Match button. If you don't, skip that address and go to the next in the list.
- Keep a log of steps you took that found successful corrections and turn this in with your assignment as **Assignment8-2YourName.docx**. For each address investigated, give the original address, its problem, source for additional information, and correction. Provide the final match statistics in the log.

## Requirements

- Create a map layout: Include a layout showing a point map of immigrant-run business locations in Pittsburgh. Symbolize the point map using type of business.

ASSIGNMENT 8-3

# Examine match option parameters for geocoding

There are no guidelines for selecting the match parameters that tune the ArcMap geocoding system. So here you'll explore the sensitivity of geocoding accuracy to variations in those parameters, and in the process make your own rough guidelines. You'll see that as match options become too lenient, ArcMap starts making obvious errors in geocoding, making matches that are clearly wrong. The data is the same used for assignment 8-2—immigrant-run businesses, including high-tech firms, restaurants, and grocery stores in Pittsburgh.

## Get set up

- Rename the folder **\EsriPress\GIST1\MyAssignments\Chapter8\Assignment8-3YourName\** to your name or student ID. Store all files that you produce for this assignment in this folder.

- Create a new map document named **Assignment8-3YourName.mxd** with relative paths. Use a state plane projection.
- Create a new file geodatabase named **Assignment8-3YourName.gdb**. Save all new files that you create in this file geodatabase.

## Build the map

Add the following spatial data to your map document:

- \EsriPress\GIST1\Data\Pittsburgh\City.gdb\ImmigrantBusinesses — sample of immigrant-run businesses in Pittsburgh. Attributes include street address, city, state, ZIP Code, and type of business (Firm, Grocery, and Restaurant).
- \EsriPress\GIST1\Data\Pittsburgh\City.gdb\PghStreets — line features of Pittsburgh street centerlines.
- \EsriPress\GIST1\Data\Pittsburgh\City.gdb\Neighborhoods — polygon features of Pittsburgh neighborhoods.

## Geocode with alternative locators

- Create a locator named **PittsburghTIGER**. Use US Address–Dual Ranges as the type of locator and PghStreets as the reference data. As address input fields, use both ADDRESS and ZIP.
- Geocode the immigrant businesses data three times with different match parameter values. Click the Geocoding Options button in the Geocode Addresses window to change parameter values.
    - Strictest: Spelling sensitivity 100, Minimum candidate score 100, Minimum match score 100.
    - Default: Spelling sensitivity 80, Minimum candidate score 10, Minimum match score 85. These are the locator's default matching parameters.
    - Most Lenient: Spelling sensitivity 0, Minimum candidate score 0, Minimum match score 0.

## Analyze results

Create a Word document, **Assignment8-3YourName.docx**. Include a title and your name. Create and fill out the following table and include it in your document:

| | Strictest | | Default | | Most Lenient | |
|---|---|---|---|---|---|---|
| | **Number** | **%** | **Number** | **%** | **Number** | **%** |
| Match | | | | | | |
| Tied | | | | | | |
| Unmatched | | | | | | |
| Total | 112 | | 112 | | 112 | |

Using your mapped points, first turn on the Strictest and Default point layers. Select Default points on the map that are not matched in the Strictest layer. Examine the selected Default records of each such point and type comparison address data for all of Pittsburgh in a table such as the one below. Include your table (and the next one) in Assignment8-3YourName.docx. The Good or Bad Match and Reason explains geocoding behavior in transitioning from the strictest to the default locator.

### Strictest to Default

| Raw Data Address | Raw Data Zip Code | Matched Address | Matched ZIP Code | Good or Bad Match and Reason |
|---|---|---|---|---|
|  |  |  |  |  |
|  |  |  |  |  |
|  |  |  |  |  |
|  |  |  |  |  |

Do the same for the Default and Most Lenient layers. Zoom in to the Central Business District neighborhood and examine only points in that neighborhood (to save time) in a table such as the following.

### Default to Most Lenient

| Raw Data Address | Raw Data Zip Code | Matched Address | Matched ZIP Code | Good or Bad Match and Reason |
|---|---|---|---|---|
|  |  |  |  |  |
|  |  |  |  |  |
|  |  |  |  |  |
|  |  |  |  |  |

Make a recommendation on which locator to use and why for the three locators that you tried. If you were to try additional locators, which ones would you try next?

# 9

# Spatial analysis

Analysis produces information from data, in which the information isn't always obvious but is there nonetheless. Spatial analysis deals with problems or behaviors that involve geographic locations. To produce sophisticated spatial information, you need to use methods or tools that process or transform data into information. This chapter has you apply several GIS tools for spatial analysis: buffers, spatial joins, intersection of map layers, and cluster analysis.

Several exercises in the chapter employ proximity analysis—determining what features are near other features. Of course, it is uniquely GIS that can perform such analysis with ease because map features have geographic coordinates. The tool for proximity analysis is a buffer, polygons that place an area around a given set of features. In another kind of problem, intersection of map layers allows you to find areas that meet criteria. In the facility location example of this chapter, you use buffers to represent criteria for locating new facilities. Areas that are the intersection of all criteria buffers are candidates for new facilities. One additional application of buffers is to estimate the falloff of facility use—public swimming pools, in this case—the farther away the potential clients are from the nearest swimming pool, the less they use it. This is an example of the gravity model as used in GIS: the farther two objects are away from each other, the less attraction between them.

Spatial joins enable you to easily aggregate point data to polygon summaries. In the public pools case, the polygons for data summaries are buffers around pools. The results of this study can help identify gaps in the coverage of the target population by the set of public swimming pools in a city.

### Learning objectives

- Buffer points for proximity analysis
- Conduct a site suitability analysis
- Use multiple ring buffers for calibrating a gravity model

## TUTORIAL 9-1

# Buffering points for proximity analysis

Drug-free school zones are areas within 1,000 feet of a school that are designed to protect students from illicit drug dealers. Federal, state, and local laws increase penalties for illegal drug dealing within such zones. For example, Pennsylvania has a mandatory two-year minimum jail or prison sentence and a maximum four-year sentence for adult drug dealers in such cases. In this tutorial, you find all 911 drug calls for service to police for incidents in drug-free school zones. The output maps and table of incidents are materials that an analyst would study further—matching crime incident records and verifying distances from school properties to determine which cases are eligible for prosecution.

## Open a map document

1   **Start ArcMap and open Tutorial9-1.mxd from the Maps folder.** This map is for the Middle Hill neighborhood of Pittsburgh, showing all schools in or near the neighborhood along with all drug calls for service to police during a certain period of time.

**2** Save the map document to the Chapter9 folder of MyExercises.

## Buffer schools

**1** On the Menu bar, click Windows > Search.

**2** In the Search window, type buffer in the search textbox and press Enter.

**3** Click Buffer (Analysis). This tool creates circular buffers (polygons) around points of a radius that you specify. It also creates buffer areas around and including the shapes of line or polygon features.

**4** Type or make selections as shown in the image, with Schools for Input Features, SchoolBuffer for Output Feature Class, and 1,000 feet for Linear unit. The option of NONE for dissolve type has the buffer tool create a separate circular buffer for each school. Each finished buffer has the school's name, so you can intersect the buffers with drug calls and the calls within a buffer get the corresponding school name as an attribute—exactly what you need.

**5** Click OK, wait for ArcMap to finish processing, and hide the Search window.

**6** Move SchoolBuffers to the top of the table of contents and change its symbology to a hollow fill. You can see that a lot of the drug calls are in drug-free zones.

The next major step is to intersect the buffers and drug calls to identify the calls within buffers. First, however, you do some setup work to clean up attribute tables to display only needed attributes.

## Turn off unneeded attributes

**1** Open Properties on the Fields tab for SchoolBuffers. There are 21 attributes, but you need only DESCR, which is actually the school name, and STUDENTS, which is enrollment level.

**2** Click the Turn Off All Fields button at the top left of the window and click DESCR and STUDENTS to turn those fields back on. Click DESCR to select it and type School for its alias in the Appearance panel. Do the same for STUDENTS, giving it the alias Enrollment.

**3** Click OK.

**4** Similarly, for Drug Calls, display only NATURE_COD, ADDRESS, and CALLDATE with the aliases Call Type, Address, and Call Date, respectively.

**5** Save your map document.

## Intersect school buffers and drug calls

**1** Click Geoprocessing > ArcToolbox.

**2** Expand Analysis Tools > Overlay, double-click the Intersect tool, and make selections or type as shown, with **SchoolBuffers** and **Drug Calls** for Features and **IntersectBuffersDrugCalls** for Output Feature Class.

**3** Click OK. Close ArcToolbox. The intersection point layer has just the Drug Call points inside school buffers.

**4** **Open the attribute table for IntersectBuffersDrugCalls.** You can see that each drug call has the name of the school corresponding to the buffer in which it lies. Also, the only attributes in the intersection table are the ones you left visible in the input map layers, and the intersection layer keeps the aliases you created for the input layers. In practice, you could export the table to Excel and give it to an analyst for further work.

IntersectBuffersDrugCalls

| School | Enrollment | FID_CADCalls | Call Type | Address | Call Date |
|--------|-----------|-------------|-----------|---------|-----------|
| WEIL ELEMENTARY | 467 | 258 | DRUGS | 350 KIRKPAT | 7/28/2012 |
| WEIL ELEMENTARY | 467 | 248 | DRUGS | 2140 HEMAN | 7/24/2012 |
| WEIL ELEMENTARY | 467 | 246 | DRUGS | 2158 HEMAN | 7/8/2012 |
| WEIL ELEMENTARY | 467 | 242 | DRUGS | 2162 HEMAN | 8/2/2012 |
| WEIL ELEMENTARY | 467 | 223 | DRUGS | 2154 CENTRE | 8/31/2012 |
| WEIL ELEMENTARY | 467 | 224 | DRUGS | 2154 CENTRE | 8/20/2012 |

**5** Close the table and save your map document.

## YOUR TURN

You can buffer lines as well as polygons. Wylie Avenue is a major street with many commercial land uses that runs through the middle of the Middle Hill neighborhood. Commercial areas are ideal for illicit drug dealing, so an application of a buffer is to see how many drug calls are associated with Wylie Avenue. Use Select By Attributes to run the query FNAME = 'Wylie' to select all segments of that street. Then run the buffer tool with Streets as the input feature (ArcMap uses only the selected streets) with a 300-foot buffer and ALL Dissolve Type (the interior of individual line segment buffers are dissolved to make one polygon for this buffer). Save the buffer as MyExercises \Chapter9\Chapter9.gdb\ WylieStreetBuffer. Turn off SchoolBuffers and IntersectBuffersDrugCalls. Symbolize the WylieStreetBuffer with a hollow fill. Save your map document.

# TUTORIAL 9-2

# Conducting a site suitability analysis

Suitability analysis for facility location is a classic GIS application that depends on two GIS functions: buffers and intersection. In this tutorial, you'll find potential areas for new police satellite stations in each car beat of Rochester, New York's Lake Precinct. A car beat is the territory of a patrol car. Criteria for locating these stations are that the site must be centrally located in each car beat (within a 0.33-mile radius buffer of car beat centroids), in retail/commercial areas (within 0.10 mile of at least one retail business), and within 0.05 mile of major streets. The suitable areas are those in the intersection of all three buffers.

## Open a map document

1   **Open Tutorial9-2.mxd from the Maps folder.** This map document contains a map of the Lake Precinct of the Rochester Police Department. It includes police car beats, car beat centroids, retail business points, and street centerlines.

2   **Save the map document to the Chapter9 folder of My Exercises.**

You'll start out by making three sets of buffers for implementing siting criteria.

# Buffer car beat centroids

**1**  **Click Windows > Search, search for buffer, and open the buffer tool.**

**2**  **Type or make selections as shown in the image, with** CarBeatCentroids **for Input Features,** CarBeatCentroidBuffers **for Output Feature Class,** 0.33 **Miles for Linear unit, and PLANAR for Method.**

**3**  **Click OK and hide the Search window.**

**4**  **Symbolize the buffers polygons with a hollow fill, an outline color of Mars Red, and an outline width of 1.**

Next, you'll need to find areas within the car beat buffers that meet the remaining criteria.

# YOUR TURN

Create two more sets of buffers with the following characteristics:

Retail businesses:
- Input Features = Retail Businesses
- Output Feature Class = RetailBusinessesBuffer (in MyExercises\Chapter9\)
- Linear unit = 0.10 miles
- Dissolve Type = ALL
- Symbolization is a hollow fill with a blue outline.

Streets:
- Use Select By Attributes to select Streets with the query condition FCC = 'A40' OR FCC = 'A41'. Use the Get Unique Values button for values of FCC.
- Input Features = Streets
- Output Feature Class = MajorStreetsBuffer
- Linear unit = 0.05 miles
- Dissolve Type = ALL
- Symbolization is a hollow fill with a green outline.

Turn off the buffer inputs, leaving just the buffers and Car Beats on, to get a look at the results thus far.

# Intersect buffers

ArcMap lets you intersect only two map layers at a time. So the approach is to intersect two buffers and then to intersect the resulting intersected layer with the third buffer.

**1**  Search for intersect **and open the Intersect (Analysis) tool.**

**2**  **Type or make selections as shown, with CarBeatCentroidBuffers and RetailBusinessesBuffer for Features,** BeatsBusinessesIntersect **for Output Feature Class, and ALL for JoinAttributes.**

**3**  Click OK.

> ## YOUR TURN
>
> Intersect BeatsBusinessesIntersect with MajorStreetsBuffer to produce SuitableSites. Symbolize the end product as you wish and turn off the original buffers and Intermediate products but turn Streets and Retail Businesses back on. The following map is zoomed in to one of the car beats to show some details. Save your map document.

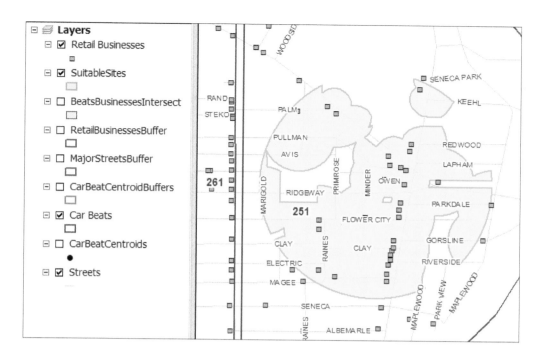

## TUTORIAL 9-3
# Using multiple ring buffers for calibrating a gravity model

The gravity model states that the farther two objects are from each other, the less the attraction between them. The falloff in attraction with distance is often nonlinear and rapid, as in Newton's gravity model in which the denominator of attraction is distance squared. Here you'll analyze the intention of youths, ages 5 to 17, to use public pools in Pittsburgh. During a budget crisis, Pittsburgh officials closed 16 of 32 public swimming pools. In the first year of the pools being closed, "pool tags," which must be worn to gain admission to a pool, were free. So, if youths intended to use a pool, all they had to do was get free pool tags. The question is whether distance from the nearest open pool had any impact on use.

A useful spinoff benefit of this case is that the available data is a random sample of pool tag holders so you get to deal with some issues associated with working with a random sample. The sample is representative of the total population but, of course, is only part of the population. For the pool case study, you need estimates for the total population of youths and therefore need to calculate a scale-up factor. There were 56,162 pool tag owners during the summer of the study year, and the sample has 2,285 geocoded residences of pool tag owners. A separate sample estimates that 46.1 percent of pool tag holders are youths, ages 5 to 17.

Thus, to scale up the sample results of your analysis to the population level, you need to multiply the number of sampled pool tag holders by a factor of 0.461 (56,162/2,285 = 11.3). The ratio in parentheses scales up to the total population of pool tag holders and the 0.461 adjusts that ratio to just youths 5 to 17. The 11.3 scale-up factor assumes that all parts of Pittsburgh have the same age distribution. which, of course, leads to some errors.

## Open a map document

**1** **Open Tutorial9-3.mxd from the Maps folder.** The data for this tutorial includes the open and closed pools, the random sample of residence addresses of pool tag owners (youths in this case), and census block data on the population of youths. You'll use a multiple-ring buffer to determine (1) the number of pool tag owners at a set of distances from the nearest open pool (scaled up from the sample by the 11.3 factor) and (2) the target youth population at the same set of distances. From this buffering, you can calculate the percentage of youths who owned pool tags at a range of distances from the nearest open pool as well as the total youth population in each range.

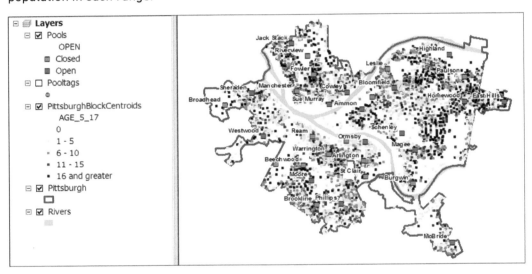

**2** **Save the map document to the Chapter9 folder of MyExercises.**

**3** **Turn Pooltags on and then back off.** Youths with pool tags live inside and outside of the city, as you can see.

You can't tell from the map whether youths closer to an open pool are more likely to have pool tags or not. You need to do some analytical work to answer that question.

# Create multiple ring buffers for open pools

You are about to use the Multiple Ring Buffer tool in ArcToolbox. First, however, you must select only open pools. Then the Multiple Ring Buffer tool creates buffers for only the open pools.

**1** Click Selection > Select By Attributes, and type or make selections as shown, with Pools for Layer, "OPEN" for Method, and "OPEN = 1" for the WHERE clause.

**2** Click OK.

**3**    Search for and open the Multiple Ring Buffer (Analysis) tool. Type or make selections as shown, with Pools for Input Features; PoolBuffers for Output Feature Class; 2, 1.5, 1, 0.75, 0.5, and 0.25 for Distances; Miles for Buffer Unit; and distance for Field Name . It doesn't matter which order you use to enter Distances. To match the figure, however, enter the "2" first and work backward to 0.25.

**4**    Click OK and close the Search window. Be patient. The Multiple Ring Buffer tool takes a minute or so to execute.

**5**  Click Selection > Clear Selected Features.

**6**  Double-click the paint chip for the 0.25 row, select No Color for fill color, select 1.50 for outline width, Tourmaline Green (column 8, row 3) for outline color, and click OK.

**7**  Similarly change each successive paint chip with no color fill and outline colors working your way left from Tourmaline Green (Medium Apple, Peridot Green, Solar Yellow, Electron Gold, and Fire Red). The image for step 5 shows the resulting map zoomed in to a part of Pittsburgh. Youths residing in areas within green to yellow lines are close to an open pool, within one mile. Those from yellow out to red are farther, from one to two miles. You can see there is a large area between orange and red lines (1.5 to 2.0 miles from the nearest pool), in which there are no closed pools to reopen and fill the gap. Unless a new pool was built, that area appears to have poor access to public pools.

9

## Spatially join buffers, pool tag owners, and youth population

You easily can use spatial joins to sum up the number of pool tag owners and youth population in each ring of the buffers, and then calculate the corresponding use rate per buffer, the number of pool tag owners divided by youth population.

**1** In the table of contents, right-click PoolBuffers, click Joins and Relates > Join, and type or make selections as shown, with "Join data from another layer based on spatial location" for "What do you want to join this layer?" and sum for "How do you want the attributes to be summarized?"

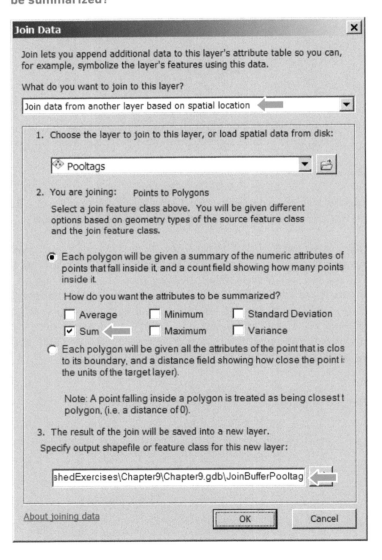

**2** Click OK.

**3**  Open the attribute table of the new layer and see that it has the attribute Count_, which is the number of youths with pool tags in each ring of the buffer. Close the table.

**4**  Right-click JoinBufferPooltag, click Joins and Relates > Join, and type or make selections as shown, with "Join data from another layer based on spatial location" for "What do you want to join to this layer?"; PittsburghBlockCentroids for "Choose the layer to join to this layer"; .Sum for "How do you want the attributes to be summarized; and JoinBufferPooltagYouth for "Specify output shapefile or feature class for this new layer."

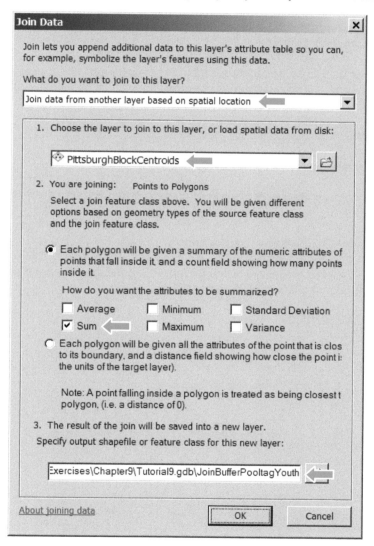

**5**  Click OK.

**6**  Open the attribute table of the new layer and see that it has the attribute Sum_AGE_5_17, which is the number of youths in each ring of the buffer. Leave the table open.

# Compute use rate

**1**   **Add a new field,** UseRate, **to the attribute table of JoinBufferPooltagYouth with the float data type.**

**2**   **Right-click UseRate and click Field Calculator.**

**3**   **Create the expression** UseRate = 100*11.3* [Count_] / [Sum_AGE_5_17]. **Click OK.** Count_ is the number of records in each buffer ring, the same thing as the number of sampled pool tag owners. The 100 makes the computed fraction a percentage and the 11.3 is the scale-up factor from sample to population. Examine UseRate. It declines with distance as expected.

You will create a scatter plot of UseRate versus Distance, but first you'll need to compute a better distance for each buffer ring. The Distance attribute is the maximum radius of each buffer ring. Better representing each buffer is its average radius, which you'll compute next.

**4**   **Add another new float data type field,** DistanceAverage.

**5**   **Click Customize > Toolbars > Editor.**

**6**   **Click the list arrow on the Editor toolbar > Start Editing, select JoinBufferPooltagYouth, and click OK > Continue.**

**7**   **Starting in the first row of the DistanceAverage column of the attribute table and working down the column, type the following values:** 0.125, 0.375, 0.625, 0.875, 1.25, 1.75. **For this** figure, columns were dragged left or right to group Distance, DistanceAverage, and UseRate so that you can see the relationship between these attributes.

| Distance | DistanceAverage | UseRate |
|---|---|---|
| 0.25 | 0.125 | 87.39027 |
| 0.5 | 0.375 | 65.25417 |
| 0.75 | 0.625 | 51.68109 |
| 1 | 0.875 | 44.07652 |
| 1.5 | 1.25 | 40.79899 |
| 2 | 1.75 | 41.04866 |

**8**   **On the Editor toolbar, click the list arrow > Save Edits. Click Stop Editing, close the Editor toolbar, and close the attribute table.**

# Create a scatter plot of use rate versus average radius for each buffer ring

**1** Click View > Graphs > Create Graph.

**2** Type or make selections as shown, with Scatter Plot for Graph type, JoinBufferPooltagYouth for Layer/Table, UseRate for Y field, Distance Average for X field, and DistanceAverage for X label field. Leave Add to legend unchecked.

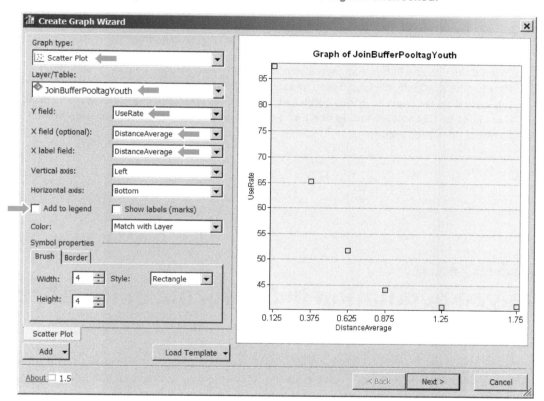

**3** **Click Next > Finish.** You can see that the relationship between use rate and average buffer radius is smooth but with rapid falloff. Within 0.125 miles of an open pool, about 88 percent of youth have pool tags, but by 1.25 miles average distance, use rate is cut by more than half to 41 percent. The relationship includes some random error about a curve that you can imagine passing smoothly through the points because of having only a random sample plus assumptions made to allow use of the 11.3 scale-up factor.

**4** Save your map document and graph to the Chapter9 folder of MyExercises.

---

**YOUR TURN**

Turn off the two JoinBuffer map layers so that you can see the pools, your original buffers, PittsburghBlockCentroids (population by block of 5- to 17-year-old youths), and the Rivers. If you were director of parks in Pittsburgh and had the budget to open one more pool and could close open pools and open closed pools on a one-for-one basis, what would you do? When finished, close your graph.

---

ASSIGNMENT 9-1

# Analyze population in California cities at risk for earthquakes

In this assignment, you'll use GIS to create buffers around major earthquakes that have occurred in California and determine how many people live in urban areas near these events. In the process, you'll learn a few new ways to work with the Buffer tool.

## Get started

First, rename your assignment folder, create a map document, and create a file geodatabase.

- Rename the folder \EsriPress\GIST1\MyAssignments\Chapter9\Assignment9-1YourName\ to your name or student ID. Store all files that you produce for this assignment in this folder.
- Create a new map document, **Assignment9-1YourName.mxd** with relative paths.
- Create a new file geodatabase, **Assignment9-1YourName.gdb**. Save all the map layers and tables you create in this file geodatabase.

# Requirements

Use a UTM projection appropriate for California in your data frame. Use this projection for all map layers that you create. Include the following map layers and data in your map document:

- \EsriPress\GIST1\Data\UnitedStates.gdb\CACounties—polygon features of California counties
- \EsriPress\GIST1\Data\UnitedStates.gdb\USCities_dtl—point features for cities in the United States. Attribute POP2007 has city populations. Records without data for this attribute have the value –99999, which stands for missing value. Use a query definition to eliminate those cities with a value of –99999. Create a new feature class, **CACities**, with only California cities and remove the original USCities_dtl layer.
- \EsriPress\GIST1\Data\DataFiles\Earthquakes.dbf—table of earthquakes in California, off the coast of California, and in southern Oregon with latitude and longitude attributes in North American Datum 1983 coordinates. Be sure to specify these coordinates, and not UTM, when using Create Feature Class From XY Table ArcCatalog. The attribute MMI—the Modified Mercalli Intensity scale, currently used in the United States—has values reflecting expected impacts. Values 7 to 12 correspond to increasing damage, from considerable damage of poorly constructed buildings to destruction.

Create a 20-mile buffer around earthquakes whose MMI is 7 or greater for 1985 through to the end of the data. Include a label with the total population within buffers. Export a layout with your map, **Assignment9-1YourName.jpg**.

It takes a couple of steps to get the desired buffers for this assignment. The map needs each separate buffer area to have a label displaying the total urban population in that area. If you were to use the ALL dissolve type when buffering earthquakes, a single polygon would result for all buffers, even though they are separate areas. Some 20-mile buffers overlap. Each set of overlapping buffers needs to be dissolved to form a single buffer polygon. Many 20-mile buffers do not overlap. These need to be separate polygons. The approach to building the needed buffers uses a dissolve field that you can create in the following steps:

- Buffer earthquakes with an MMI greater than or equal to 7 from 1985 and later using a 20-mile radius, the NONE dissolve type, and saved to the feature class **Earthquakes_Buffer**. Do not use the dissolve fields in the buffer tool, although you must do some dissolving in this assignment later. Several circular buffers overlap.
- Open the attribute table of Earthquakes_Buffer and add a new field named **BufferGroup** with the Short Integer data type.
- Work from north to south. Use the Select Features tool to select the overlapping buffers along the north coast of California. Then, using the Field Calculator or Editor toolbar, set the members of the selected buffers to the value 1 in the BufferGroup field. Repeat this step for the next set of overlapping buffers but use the value 2 for BufferGroup. Finally, give the remaining two groups of overlapping buffers unique values for BufferGroup.
- Use the Dissolve tool to dissolve Earthquakes_Buffer using BufferGroup as the dissolve field. Don't use the statistics fields in the Dissolve tool. Name the new feature class **Buffer_Earthquakes2** and save it to your file geodatabase. The result is separate polygon buffers for each nonoverlapping buffer, plus polygons for each set of overlapping buffers.

- Spatially join CACities to Buffer_Earthquakes2 to create **Join_CitiesBuffer**. Start the join by right-clicking Buffer_Earthquakes2 and be sure to use SUM so that city attributes are summed by buffer polygon for labeling your map. Use SUM_POP_2007 to label the buffers.
- Rename and symbolize the layers in the table of contents.

## ASSIGNMENT 9-2

# Analyze visits to the Jack Stack public pool in Pittsburgh

Tutorial 9-3 had you analyze intentions of youths (or their parents/guardians) to use Pittsburgh public pools during a specific summer. Intention was represented by owning a pool tag needed for admission to a pool. You found that the intended use rate, calculated as the percentage of youths who owned pool tags, fell off quickly with distance from the nearest open pool. In this assignment, you'll study visits of youths to a public pool in Pittsburgh named Jack Stack. A limitation of your study is that you do not determine how many pool tag owners near enough to Jack Stack pool never visited it. You analyze only pool tag owners who visited the pool at least once. (Although not a requirement of this assignment, you could analyze the pool tag data and see if you can determine how many pool tag owners near Jack Stack never visited the pool. You'd find that most visited at least once.)

## Get started

First, rename your assignment folder and create a map document.

- Rename the folder **\EsriPress\GIST1\MyAssignments\Chapter9\Assignment9-2YourName\** to your name or student ID. Store all files that you produce for this assignment in this folder.
- As a shortcut to building your map document, open Tutorial9-3 from the Maps folder and save it as \EsriPress\GIST1\MyAssignments\Chapter9\Assignment9-2YourName\**Assignment9-2YourName.mxd** where you use your name or student ID. Remove the Pooltags layer from the map document.

# Process the data

Add \EsriPress\GIST1\Data\Pittsburgh\PublicPools.gdb\**JackStackVisits** to your map document. This is data from a random sample of visitors to Jack Stack pool. Attributes include:

- POOL = name of the pool that the pool tag owner stated that he/she intended to use
- NOPERSONS = number of persons in the party or family for a visit
- AGE = age of the person in the party included in the random sample
- VISITS = number of visits that the person included in the random sample made to the pool during the summer studied

# Requirements

- Add the following definition query to JackStackVisits so that you analyze only youths: "AGE" <=17 AND "AGE" >= 5.
- Create a multiple ring buffer layer for the Jack Stack pool named **JackStackBuffer** using the same buffer distances as in Tutorial9-3: 0.25, 0.5, 0.75, 1, 1.5, and 2 miles. **Hint:** Select Jack Stack pool in the Pools layer to restrict the buffer to that pool. Intersect the buffer with Pittsburgh in a layer named **IntersectPittsburghBuffer** and use this intersection layer or further joins to it for display and spatial joins. You need to exclude pool users residing outside Pittsburgh, and the intersection accomplishes that. Although non-Pittsburgh youths are welcome to use Pittsburgh public pools, as a matter of policy the pool system evaluation is limited to Pittsburgh residents.
- Compute the following average for VISITS for each buffer ring. **Hint:** Right-click IntersectPittsburghBuffer, click Joins and Relates > Join, select Join data from another layer based on spatial location, join JackStackVisits, and select Average. Name this feature class **JoinBufferAverageVisits**.
- Add an attribute, named **DistanceAverage**, to JoinBufferAverageVisits for average distance of buffers as done in tutorial 9-3. Use the Edit toolbar, edit the attribute table, and type the distances **0.125**, **0.375**, **0.625**, **0.875**, **1.25**, and **1.75**, starting in the top row. There is no need to use the 11.3 scale-up factor from tutorial 9-3. The average that you calculate is representative of the population.
- Create a scatter plot of average number of visits by youth versus average distance to buffers and include it in a layout with your map. Add the graph to your layout by right-clicking it. Although the relationship has more scatter than the one you found for intentions, a rapid decline is evident in average visits with increasing distance from the pool.

# 10

# ArcGIS 3D Analyst for Desktop

This chapter is an introduction to the ArcGIS® 3D Analyst™ extension for ArcGIS Desktop software that enables 3D display and processing of maps. 3D viewing can provide insights that would not be readily apparent from a 2D map of the same data. For example, instead of inferring the presence of a valley from 2D contours, in 3D you see the valley and the difference in height between the valley floor and a ridge. This chapter uses topography, curb, and building data from the City of Pittsburgh's Mount Washington and Central Business District neighborhoods to show you how to display and analyze data in 3D. It also introduces ArcGlobe, a 3D visualization application that includes rich basemaps.

Learning objectives

- Create 3D scenes
- Create a triangulated irregular network (TIN)
- Drape features over a TIN
- Navigate through 3D scenes
- Create a 3D animation
- Use 3D effects
- Edit 3D objects
- Conduct landform analysis
- Explore ArcGlobe

## TUTORIAL 10-1
# Creating a 3D scene

ArcGIS 3D Analyst is one of a collection of extensions for ArcGIS Desktop software. You must have the extension installed on your computer and then enable it in ArcMap, which you will do next.

## Add the 3D Analyst extension and toolbar

**1**  Start ArcMap with a new blank map.

**2**  On the Menu Bar, click Customize > Extensions.

**3**  Select the check box next to the 3D Analyst extension and click Close.

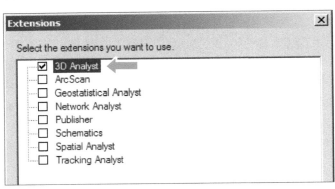

**4**  Click Customize > Toolbars > 3D Analyst. The toolbar for 3D Analyst appears. You can dock it with other toolbars below the main menu.

## Launch ArcScene and add a topography layer

ArcMap is for 2D maps, so to work with 3D maps, you need a new package. It's called ArcScene.

**1**  On the 3D Analyst toolbar, click the ArcScene button.

**2**  Click New Scenes > Blank Scene and click OK. A new, untitled scene window opens.

**3**  In ArcScene, click the Add Data button, browse through the Data folder to 3DAnalyst.gdb, and click Topo > Add. A topography layer of contours near downtown Pittsburgh appears as a 3D view, although the display is 2D at the moment.

You'll make it 3D next.

## Set document properties and save the 3D scene

**1**  Click File > Scene Document Properties.

**2**  Click the Pathnames Option button to store relative paths and click OK.

**3**  Click File > Save As, navigate to the Chapter10 folder of MyExercises, and save the 3D scene as Tutorial10-1.sxd. ArcScene scenes have a different file type—SXD—from map documents.

## TUTORIAL 10-2
# Creating a TIN from contours

3D uses the TIN representation for modeling surfaces. TIN is a vector data model of contiguous, nonoverlapping triangles sharing edges with vertices created from adjacent sample points of x, y, and z values from 3D space. Next, you'll create a TIN from the topography contour map.

# Create a TIN

**1**  **Save the scene as** Tutorial10-2.sxd **to the Chapter10 folder.**

**2**  **Click the ArcToolbox button, click the plus sign (+) next to 3D Analyst Tools, click the plus sign next to Data Management, and click the plus sign next to TIN.**

**3**  **Double-click Create TIN.**

**4**  **In the Create TIN window, save the output TIN as** chapter10_tin **to the Chapter10 folder of MyExercises.**

**5**  **Set Coordinate System to NAD_1983_StatePlane_Pennsylvania_South_FIPS_3702_Feet.**

**6** Select Topo as the input feature class and Hard_Line as the SF Type.

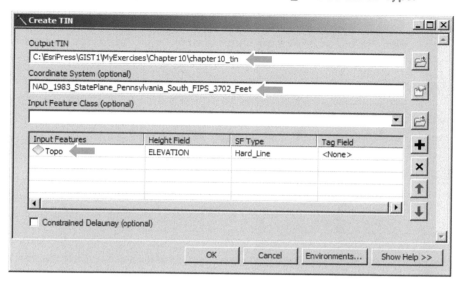

**7** Click OK. Wait while ArcGIS 3D Analyst creates a TIN from the topography contour lines and adds it as a new layer.

**8** Close ArcToolbox. The resulting map shows both the original topography (2D) and the TIN (3D).

# Change the TIN appearance

**1** In the table of contents, right-click chapter10_tin and click Properties.

**2** Click the **Symbology** tab and clear the check boxes next to **Edge Types** and **Elevation** to turn them off.

**3** Click the **Add** button, click **"Faces with the same symbol,"** click **Add,** and close the **Add Renderer** window.

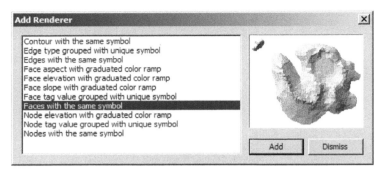

**4** Click the **Symbol** color for **Faces,** click **Early Desert Tan,** and click **OK.** This creates a more realistic-looking surface map.

**5** Click **Apply** and click **OK.** The topography layer is now one color.

## Navigate a 3D view

You do not have a good view of the TIN yet, but you can get one with some navigation.

**1   On the Tools toolbar, click the Navigate button.**

**2   Click and drag the map to view the scene from different angles.** You see that Pittsburgh is fairly flat where the three rivers converge (this area is known as "the Point") and hilly in the Mount Washington neighborhood to the right of the Point in the image.

## Zoom to see TIN's triangles

Zooming to a small area allows you to see the triangulated irregular network.

**1**  Click the Zoom In button.

**2**  Zoom to a small area on Mount Washington. There are many small triangular facets making up the surface of the chapter10_tin.

**3**  Click the Full Extent button.

**4**  Save the scene.

## TUTORIAL 10-3
# Draping features onto a TIN

Now that you have a 3D TIN, you can drape 2D map layers on it to show them in 3D.

## Drape curbs

**1**  Save the scene as Tutorial10-3.sxd to the Chapter10 folder.

**2**   In the table of contents, turn the original Topo layer off.

**3**   Click the Add Data button, browse to 3DAnalyst.gdb, and click Curbs > Add. Curbs appear below the TIN contours.

**4**   In the table of contents, right-click the Curbs layer and click Properties.

**5**   Click the Base Heights tab and select the radio button next to Floating on a custom surface. The base height is the elevation at which this flat map layer appears.

**6**   Type 10 as the Layer offset number. This offset elevates the sidewalks slightly so they do not "bleed" into the contours.

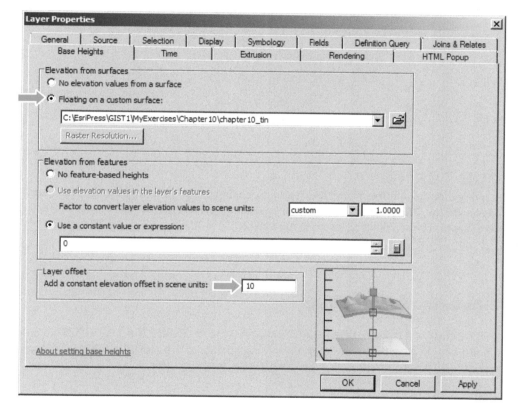

**7**   Click Apply and click OK.

**8** **Change the color of the Curbs layer to medium/dark gray.** The resulting map displays the curbs draped over contours.

## YOUR TURN

Add rivers polygon features from 3DAnalyst.gdb, change the color to Water Body blue, and drape it over the TIN. Set the Layer offset to 3. Use the Navigate tool to view the scene from different angles. When finished, zoom to full extent.

## Drape and extrude buildings

**1** **Click the Add Data button, browse to 3DAnalyst.gdb, and click Bldgs > Add.** This action adds the buildings layer with the number of building stories in the attribute table. The number of stories for Pittsburgh buildings is fictitious but a good estimate.

**2** **Right-click the Bldgs layer and click Properties.**

**3** **Click the Base Heights tab, select the radio button next to Floating on a custom surface, and choose chapter10_tin.**

**4** **Type 5 for the Layer offset.**

**5** **Click the Extrusion tab and select the check box next to Extrude features in layer.**

**6**  Enter the following expression for the Extrusion value: **[STORIES] * 10.** The building height is approximated assuming that each story is 10 feet high. You can use the calculator expression button to choose the field.

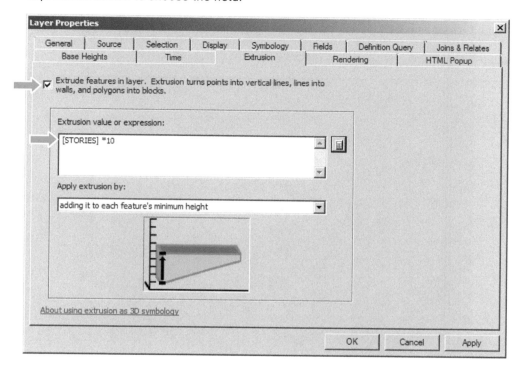

**7**  Click OK.

**8**  Change the color of the buildings to Sunset Orange. The resulting view is buildings with various heights all positioned just above the TIN. Most of the buildings in the Mount Washington neighborhood are residential houses, whereas the downtown area contains high-rises.

# Drape an aerial image

In addition to vector features, you can drape images onto a TIN. In this exercise, you will drape a USGS aerial image downloaded as a National Agriculture Imagery Program (four band) file. A better-quality image also could be downloaded but would be a much larger file size.

**1**  Turn the Bldgs layer off and navigate to a view like that in the image.

**2**  Click the Add Data button, browse to 3DAnalyst.gdb, and click Image > Add. This action adds the USGS aerial image.

**3**  Right-click the Image layer, click Properties > Base Heights tab, and make the selections as shown, with "Floating on a custom surface" marked, 0.9900 for Factor to convert layer, and 20 for Layer offset. and click OK. The USGS image uses UTM coordinates whose base unit is a meter, so some conversions are necessary. The USGS image will be draped over the TIN.

10

> **YOUR TURN**
>
> Practice adjusting the base height properties of the USGS image. Save your scene.

## TUTORIAL 10-4
# Navigating scenes

Navigation in 3D is more complicated than in 2D, so ArcScene adds additional navigation tools.

## Center view on target location and set observer

**1** Save the scene as Tutorial10-4.sxd to the **Chapter10 folder**.

**2** Remove the Image layer, turn Bldgs on, and zoom to the map extents.

**3** On the Tools toolbar, click the Center on Target button.

**4** Click a location in the Mount Washington neighborhood that overlooks the city of Pittsburgh.

**5**  On the Tools toolbar, click the Set Observer button.

**6**  Click a location in which the three rivers meet, called the Point, to set the observer location. See the top image. The resulting view (bottom image) is from the perspective of an observer at the Point, looking toward the Mount Washington neighborhood.

> ## YOUR TURN
> Practice using the Zoom To Target tool, which zooms to the target location.

## Fly through a scene

**1** Zoom to the full extent.

**2** Set the observer location to the confluence of the three rivers at the Point. The scene smoothly shifts to that location.

**3** From the Tools toolbar, click the Fly button.

Get ready for a wild flight!

**4** Click anywhere in the scene with the bird cursor and click again to start your flight.

**5** Slowly move the mouse to the left, right, up, or down. Click the left mouse button to increase your speed, and click the right mouse button to decrease your speed.

**6** Press ESC on the keyboard to stop the flight. A new view appears where you stopped the fly-through.

**7**   Zoom to the full extent.

## Create multiple views

3D views can be complicated. Using multiple views allows you to zoom to small areas when still zooming to the overall map.

**1**   From the Standard toolbar, click the Add Viewer button  .

**2**   With the Viewer1 window selected, click the Target, Observer, Navigate, and Zoom buttons to change the view. You also can use the scroll wheel on your mouse to zoom in and out.

10

**3**  Save the scene.

# Creating an animation

3D animations allow you to record movements within your views so that you can save and play them back later.

## Add the animation toolbar

**1**  Save the scene as Tutorial10-5.sxd to the Chapter10 folder.

**2**  Close the Viewer1 window.

**3**  Right-click anywhere in the blank area of a toolbar.

**4**  Click Animation to display this toolbar. The Animation toolbar appears.

**5**  Zoom to the full extent.

## Record and play an animation

**1**  Click the Open Animation Controls button on the Animation toolbar.

**2**  Click the Record button ⬤.

**3**  Click the Fly button and create a fly-through anywhere in your scene, and then click the ESC key to end your flight.

**4**  Click the Stop button ■.

**5**  Click the Play button ▶.

## Save an animation

**1**  On the Animation toolbar, click Animation > Save Animation File.

**2** Save the animation as Tutorial10-5 to the Chapter10 folder. This saves the animation as an ArcScene (.asa) file.

## Load an animation

**1** On the Animation toolbar, click Animation > Load Animation File.

**2** On the Open Animation dialog box, navigate to your Chapter10 folder and click Tutorial10-5. asa > Open.

**3** Click the Play button on the Animation Controls toolbar.

**4** Zoom to the full extent.

> ## YOUR TURN
> Practice creating animations by zooming in to a small area first. Explore the Options menu on the Animation Controls toolbar to see animation play and restore options.

## Export an animation to video

If you want to share your animation with someone who does not have ArcScene, you can export it in a common format such as an .avi file.

**1** On the Animation toolbar, click Animation > Export Animation.

**2** Navigate to the Chapter10 folder, save your animation as Tutorial10-5.avi, and then click Export and click OK. Wait until the animation is fully exported.

**3** Close the Animation Controls and Animation toolbars and save your scene.

> ## YOUR TURN
> Launch a video player such as Windows Media Player and play the AVI video that you created. If you have trouble loading and playing your video, choose the animation file in the Chapter10 folder.

## TUTORIAL 10-6
# Using 3D effects

Special effects such as transparencies, lighting, and shading modes can enhance the 3D experience for the viewer. For example, a city planner might want to differentiate between existing and proposed buildings by making one or the other transparent. In this tutorial, you'll select buildings in one neighborhood to make them transparent and add proposed buildings as solid polygons.

The Layer Face Culling command turns off the display of front or back faces of an aerial feature or graphic. Layer Lighting turns lighting on or off for the selected layer. Shading Mode allows you to define the type of shading (smooth or flat) to use for the layer selected. Depth Priority allows you to define which 3D layer should be given higher priority. This is useful when you have two 3D polygon layers that share the same location and might obstruct each other, such as land parcels and buildings.

## Prepare a study area

**1** Save your scene as Tutorial10-6.sxd **to the Chapter10 folder and zoom to full extent.**

**2** **Choose Selection > Select By Attributes and build the query as shown, with Bldgs for Layer:** "NEIBHBORHOOD" = 'North Shore'. **Click OK.** This selects buildings on the North Shore neighborhood.

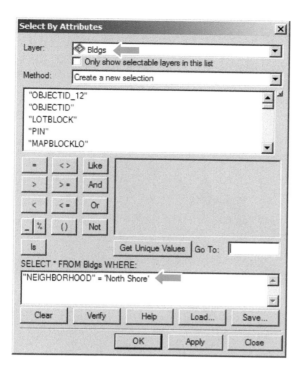

**3** In the table of contents, right-click Bldgs and click Selection > Create Layer from Selected Features.

**4** Rename the layer North Shore Bldgs and clear the selected features. Turn off the original Bldgs layer.

**5** In the table of contents, right-click North Shore Bldgs and click Properties > the Base Heights tab.

**6** Click "Floating on a custom surface" and type 10 for Layer offset.

**7** Click the Extrusion tab, select the check box next to Extrude features in a layer, enter the expression [STORIES] *10, and click OK.

**8** Change the North Shore Bldgs color to Ultra Blue.

**9** Turn off the original Bldgs layer and zoom to the North Shore Bldgs layer. The 3D scene shows the existing buildings on the North Shore.

10

**YOUR TURN**

Add the layer from 3DAnalyst.gdb named ProposedBldgs, drape them to the TIN, extrude using [STORIES] * 10, an offset from surface of 10, and Mars Red color.

## Create a transparency effect

**1**  Right-click anywhere in the blank area of a toolbar and click 3D Effects to display the 3D Effects toolbar.

**2**  On the 3D Effects toolbar, click the Layer drop-down menu and click North Shore Bldgs. Click the Layer Transparency button.

**3**  Drag the slider to change the layer's transparency to 75%.

**4**  On the 3D Effects toolbar, click the Layer Face Culling button ⬧, and select the option next to Honor features built-in face culling. The resulting effect is a nice transparency for existing buildings on the North Shore to better focus on the proposed buildings.

## YOUR TURN

Experiment with changing other effects for your 3D buildings, including Layer Lighting, Shading Mode, and Depth Priority. Change the effects for the Rivers and Curbs layers in your 3D scene. Save your scene.

## TUTORIAL 10-7

# Using 3D symbols

The 3D Analyst extension comes with many 3D symbols for objects such as trees, which you will use next.

## Add trees layer

1  **Save your scene as** Tutorial10-7.sxd **to the Chapter10 folder.**

2  **Click the Add Data button.**

3  **In the Add Data browser, browse to 3DAnalyst.gdb and click NorthShoreTrees > Add.**

## Display points as 3D trees

**1**  In the table of contents, right-click the NorthShoreTrees layer and click Properties.

**2**  Click the Base Heights tab and select the option next to Floating on a custom surface.

**3**  Click the Symbology tab and click the Symbol button.

**4**  In the Symbol Selector window, click the Style References button (in the lower right of the window), select 3D Trees, and click OK > OK.

**5**  Click Unique Values under Categories, click TYPE for the value field, click Add All Values, and double-click the point symbol for American Elm 1.

**6**  Scroll through the symbols until you see the 3D trees, choose American Elm 1, and click OK.

**7** Double-click the symbols for the remaining trees and assign the corresponding 3D symbol. 3D symbols can be assigned for all trees except "stumps."

**8** Click OK to close the Layer Properties window.

**9** Click the Navigate > Zoom > Pan buttons to view the trees from street level. The image shows trees along a few streets in Pittsburgh's North Shore.

**10** Save your ArcScene.

---

### YOUR TURN

Add the point feature class NorthShoreVehicles from 3DAnalyst.gdb. Set the base height for the surface to chapter10_tin, offset the points 5 units, zoom to the layer, and then to a few vehicles. Display the vehicles as 3D vehicle symbols using Unique Values based on the Type value field from the feature attribute table. In the Symbol Selector window, rotate each vehicle 115 degrees by typing a new angle. Explore the other 3D symbols that come standard with ArcEditor, including 3D Basic, 3D Billboards, 3D Buildings, 3D Industrial, 3D Residential, and 3D Street Furniture. Save the scene when you are finished.

## TUTORIAL 10-8
# Editing 3D objects

You can edit 3D objects using the 3D Editor toolbar. Edits include changing 3D heights, moving 3D objects, or creating new features. In this exercise, you'll export the proposed North Shore buildings as a new feature class to edit.

## Export buildings

**1** Save your scene as Tutorial10-8.sxd **to the Chapter10 folder and zoom to the NorthShoreBldgs layer.**

**2** In the table of contents, right-click the ProposedBldgs layer and click Data > Export.

**3** Click "Use the same coordinate system as the data frame" and save this as a feature class, ProposedBldgsEdits.

**4** Add the new layer to the scene, drape it to the TIN, extrude using [STORIES * 10], and change the color to Midday Yellow.

**5** Turn the original ProposedBldgs layer off.

## Use the 3D Editor toolbar

**1**  Click Customize > Toolbars > 3D Editor.

**2**  On the 3D Editor toolbar, click 3D Editor > Start Editing > ProposedBldgsEdits and click OK.

## Edit 3D building height

**1**  On the 3D Editor toolbar, click the Edit Vertex tool.

**2**  Click one of the ProposedBldgsEdits polygons to select it.

**3**  On the 3D Editor toolbar, click the Attributes button.

**4**  In the Attributes window, type 40 for the number of stories and press Enter.

ArcScene changes the building to reflect the new height.

## Move a 3D building

**1**   On the 3D Editor toolbar, click the Edit Placement tool.

**2**   **Drag a building to a new location on the map.** The building is in a new location and drapes to the contours that it was moved to.

10

**YOUR TURN**

Practice moving and editing other proposed building heights. When finished, click 3D Editor > Stop Editing > Save Edits. Save your scene and exit ArcScene.

## TUTORIAL 10-9

# Using 3D Analyst for landform analysis

In this tutorial, you'll explore how to create steepest paths, contour profiles, and line-of-sight analysis using a TIN in ArcMap.

## Start a map document

**1**  Start ArcMap and open a new blank map.

**2**  Save the map document as Tutorial10-9.mxd to the Chapter10 folder.

**3**  If it is not already open, click Customize > Toolbars > 3DAnalyst to load the toolbar.

**4**  Click the Add Data button, navigate to the Chapter10 folder, click chapter10_tin > Add. The TIN that you created in Tutorial10-2 is added.

# Create a steepest path line and elevation profile

The Create Steepest Path tool calculates the direction a ball would take if released at a given point on the surface. The tool works with raster, TIN, and other surfaces. The result is a 3D graphic line added to the map or scene. You can then create a profile graph of this 3D graphic line.

**1** On the 3D Analyst toolbar, click the Create Steepest Path button.

**2** Click a location from the South Shore to Mount Washington as shown in the image.

**3** On the Tools toolbar, click the Select Elements button, click the lower-left grip handle, and drag to create a longer path line as shown.

**4** On the 3D Analyst toolbar, click the Profile Graph button.

**5** Double-click the profile graph title and, in the resulting window, click the Appearance tab.

**6** Type Topography Analysis **as the Graph profile title,** Change in Elevation from South Shore to Mt Washington **as the footer, and click OK.** The graph now reflects the profile analysis.

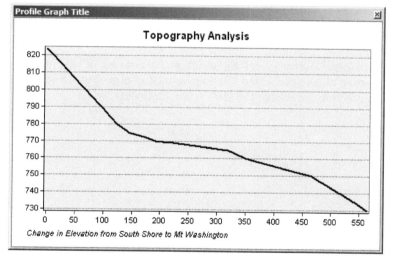

**7** Close the profile graph window.

# Create a line of sight

A line-of-sight analysis creates a graphic line between two points showing where the view is obstructed between those points.

**1**    On the 3D Analyst toolbar, click the Create Line of Sight button.

**2**    In the Line Of Sight window, type 6 as the observer and target offsets. The offset is the number of feet above the TIN for the observer and target. If you set the observer and target heights to zero, then typically you will have a view with more obstructions than one with a height greater than one.

**3**    Click a point near Pittsburgh's Point where the three rivers meet.

**4**    Click a location on Mount Washington where the elevation is above 910 feet (gray area on the map). The resulting map shows red along the line where the observer's line of sight is obstructed and green along the line where the view is not obstructed.

10

## YOUR TURN

Choose additional observer and target points to see line-of-sight visibility. Change the observer and target heights to see if the visibility changes. Create new steepest line paths and profiles. Explore other 3D Analyst functions in ArcToolbox. Click 3D Analyst Tools, Triangulated Surface to create surface aspect, contours, difference, and slope analysis. Save your map document and exit ArcMap.

## TUTORIAL 10-10
# Exploring ArcGlobe

ArcGlobe provides a seamless basemap infrastructure in an ArcScene-like interface for the entire world with imagery, elevation, political boundaries, and highways. You can add your own layers to ArcGlobe and quickly have an impressive GIS application. You need a broadband Internet connection to use the web service that provides the basemaps.

## Launch ArcGlobe

**1    On your desktop, click Start > ArcGIS > ArcGlobe 10.8.x.** ArcGlobe opens showing default layers provided by Esri.

**2    Save the document as** Tutorial10-10.3dd **to your Chapter10 folder.** ArcGlobe documents have the extension .3dd.

## Explore ArcGlobe

By default, ArcGlobe opened with the Navigate tool selected.

**1    Click, hold, and drag the display to the right so that the West Coast of North America is at the center of the map.**

**2** Place your cursor as shown in the following image.

**3** Right-click and drag downward until San Francisco appears, and then recenter the map on San Francisco using your left mouse button. If you don't know where San Francisco is located, you can add the CACounties layer from UnitedStates.gdb, select San Francisco County, and zoom to selected features.

**4** Keep this process up until you can see San Francisco and the Bay Bridge. If you need to zoom out at some point, right-click the map and drag upward. You also can use your mouse wheel to zoom in and out.

10

**5**   **Zoom in even closer as indicated in the image.** Here you can see the shadow of the Bay Bridge.

**6**   **Click Bookmarks > Create Bookmark and create a bookmark named** San Francisco Bay Bridge.

## YOUR TURN

Zoom in to a mountainous area of the globe and use the Navigation Mode button to see elevation. You can view mountains in 3D. You can find mountains using Find with the Locations tab, zooming in to them and then zooming out. Then, to view elevation, click the Navigation Mode button. For instance, the image is a view of Pikes Peak, Colorado.

## Add and display large-scale vector data

You can add and display map layers for anywhere in the world. Next, you'll add two layers for New York City.

**1** Click the Add Data button, browse through the Data folder to NYC.gdb, and click Boroughs > Add > Next > Finish > Close.

**2** In the table of contents, right-click Boroughs, click Properties > Symbology tab, and change the layer's symbol to no fill and a bright-yellow outline with width 1.

**3** In the table of contents, right-click Boroughs > Zoom To Layer.

**4** In the table of contents, drag Boroughs above the Imagery layer. You can see the New York City boroughs in yellow.

**5** Add the layer ManhattanWaterfrontParks from NYC.gdb and symbolize as a green fill.

**6** Zoom to the area of lower Manhattan.

**7** Turn the Boroughs layer off. You can see the New York City waterfront parks as polygons and the aerial image.

**8**   **Create a spatial bookmark named** Lower Manhattan.

> ## YOUR TURN
> **Practice adding other NYC layers and zoom to various boroughs in New York. Save your ArcGlobe document and close ArcGlobe.**

## ASSIGNMENT 10-1

# Develop a 3D presentation for downtown historic sites

Many US cities, including Pittsburgh, are experiencing a surge of downtown revitalization. In Pittsburgh, new condominium and apartment projects are in progress, and the city planning department wants to verify that this new development does not interfere with existing historic sites. In this assignment, you help the city planning department raise the awareness of historic sites in downtown Pittsburgh by developing a 3D model and animation of these areas.

## Get set up

- Rename the folder **\EsriPress\GIST1\MyAssignments\Chapter10\Assignment10-1YourName\** to your name or student ID. Store all files that you produce for this assignment in this folder.
- Create a new map document named **Assignment10-1YourName.mxd** with relative paths.
- Create a new file geodatabase named **Assignment10-1YourName.gdb**.
- Create a new 3D scene named **Assignment10-1YourName.sxd** with relative paths.

## Build the map

Import the following into your file geodatabase:

- \EsriPress\GIST1\Data\Pittsburgh\CBD.gdb\Histsites — polygon features of historic sites in the City of Pittsburgh's Central Business District.
- \EsriPress\GIST1\Data\3DAnalyst.gdb\Bldgs — polygon features of buildings in downtown Pittsburgh.
- \ EsriPress\GIST1\Data\3DAnalyst.gdb\Curbs — line features of curbs (sidewalks) in downtown Pittsburgh.

- \ EsriPress\GIST1\Data\3DAnalyst.gdb\Topo — line features of topography contours in downtown Pittsburgh.
- \ EsriPress\GIST1\Data\3DAnalyst.gdb\Rivers — polygon features of rivers in downtown Pittsburgh.

## Process the data

In ArcMap, add Bldgs and Histsites, and create a new feature class of buildings that have their centroids in historic sites and another of buildings whose centroids are not within historic sites. *Hint:* First select buildings in historic sites, and then use Switch Selection in the attribute table to select the nonhistoric site buildings. Save the new features, **HistSitesBldgs** and **NonHistSitesBldgs**, to your file geodatabase. Remove the original Bldgs layer from the map document.

## Requirements

- In ArcScene, add all the features in your file geodatabase.
- Create a new TIN from the Topo layer named **Assignment10-1YourName_tin** and assign it spatial reference NAD_1983_StatePlane_Pennsylvania_South_FIPS_3702_Feet.
- Remove the original Topo layer and display the TIN as "Faces with the same symbol" (Early Desert color) and no edges or elevation.
- Drape all features to the TIN using a layer offset of 10 for all layers except Rivers. Use a layer offset of 5 for Rivers.
- Extrude the buildings using the stories field multiplied by 10.
- Show the nonhistoric site buildings using a transparency effect of 60% and the historic site buildings as opaque (0% transparency).

## Create a PowerPoint presentation

Create a PowerPoint presentation named **Assignment10-1YourName.pptx** that includes the following:

- Title slide with title Downtown Pittsburgh Historic Sites Building Study, including your name and today's date
- Slides with 2D images inserted of the following 3D views:
  - Map extent
  - Map zoomed to historic sites building layer
  - Three additional views of historic sites using target, observer, and various zooms

1
2
3
4
5
6
7
8
9
10
11
12

## ASSIGNMENT 10-2
# Topographic site analysis

Pittsburgh's Phipps Conservatory (https://phipps.conservatory.org) was built in 1893 by Henry Phipps as a gift to the City of Pittsburgh. The conservatory recently underwent a major renovation with the addition of a 10,885-square-foot green-engineered welcome center, a state-of-the-art production greenhouse, and a one-of-a-kind tropical forest. In 2009, Phipps Conservatory was the site of the welcome dinner for the G20 summit where it hosted an array of world leaders. Near this site, Carnegie Mellon University is expanding and needs to perform line-of-sight analysis of the topography. The 3D Analyst extension is useful for envisioning the area and conducting important site analysis. In this assignment, you use the extension to view the topography of the area, create a 3D TIN, perform a line-of-sight analysis, and conduct slope studies.

## Get set up

- Rename the folder **\EsriPress\GIST1\MyAssignments\Chapter10\Assignment10-2YourName\** to your name or student ID. Store all files that you produce for this assignment in this folder.
- Create a new map document named **Assignment10-2YourName.mxd** with relative paths.
- Create a new file geodatabase named **Assignment10-2YourName.gdb**.

## Build the map

Import the following into your file geodatabase:

- \EsriPress\GIST1\Data\Pittsburgh\Phipps.gdb\Bldgs — polygon features buildings in the Phipps Conservatory study area.
- \EsriPress \GIST1\Data\Pittsburgh\Phipps.gdb\Topo — line features topography contours in the Phipps Conservatory study area.
- \EsriPress \GIST1\Data\Pittsburgh\Phipps.gdb\PhippsAddition — polygon feature of the Phipps Conservatory welcome center addition.

## Requirements

- In ArcMap, add the Bldgs, Topo, and PhippsAddition layers. Display Bldgs as a white fill, black outline, and labeled using the NAME field.
- Display the Phipps addition as Mars Red fill with a black outline.
- Create a TIN from the Topo layer named **Assignment10-2YourName_tin** and assign it spatial reference NAD_1983_StatePlane_Pennsylvania_South_FIPS_3702_Feet.

- Create a 36-by-24-inch map layout with three data frames. Include the buildings, Phipps addition, and TIN in all three data frames.
- Rename data 1 **Slope Analysis**, remove the original Topo layer, and display the TIN as "Face slope with a graduated color ramp." Remove Edge Types and Elevation from the symbology and display the slope as a light-green to dark-green color ramp with nine quartile classes.
- Rename data frame 2 **Line of Sight Analysis**, display the TIN as "Faces with the same symbol" and a light-yellow color. Remove the other Edge Types and Elevation from the symbology. Create a line-of-sight analysis using 6 feet as the observer and target offsets from the Phipps addition to Café Phipps and CMU's Scaife Hall. Create a line-of-sight analysis from Scaife Hall to the Boundary Street building.
- Rename data frame 3 **Topography Analysis**, display the TIN as "Faces with the same symbol" and a light-yellow color. Remove the other Edge Types and Elevation from the symbology. Zoom to buildings Scaife Hall and Boundary Street building and create a 3D line from Scaife Hall to the Boundary Street building. This hillside is the location of future CMU building development. Create a 3D profile graph with the title **Topography Analysis** and footer subtitle **Change in elevation from Scaife Hall to Boundary Street**. Add the 3D graph to your layout, under data frame 3.
- Add titles, scale bars, a legend for data frame 1, and other elements you think necessary for your layout and export it as **Assignment10-2YourName.jpg**.

## ASSIGNMENT 10-3
# 3D animation of a conservatory study area

## Get set up

- Rename the folder **\EsriPress\GIST1\MyAssignments\Chapter10\Assignment10-3YourName\** to your name or student ID. Store all files that you produce for this assignment in this folder.
- Create a new map document named **Assignment10-3YourName.mxd** with relative paths.
- Create a new file geodatabase named **Assignment10-3YourName.gdb**.
- Create a new 3D scene named **Assignment10-3YourName.sxd** with relative paths.

## Build the 3D GIS

Import the following into your file geodatabase:

- \EsriPress\GIST1\Data\Pittsburgh\Phipps.gdb\Bldgs — polygon features of the buildings in the Phipps Conservatory study area.
- \EsriPress \GIST1\Data\Pittsburgh\Phipps.gdb\PhippsAddition — polygon features of the Phipps Conservatory welcome center addition.

- \EsriPress \GIST1\Data\Pittsburgh\Phipps.gdb\Curbs — line features of sidewalk curbs in the Phipps Conservatory study area.
- \EsriPress \GIST1\Data\Pittsburgh\Phipps.gdb\Topo — line features of topography contours in the Phipps Conservatory study area.
- \EsriPress \GIST1\Data\Pittsburgh\Phipps.gdb\Image — USGS ortho image in the Phipps Conservatory study area.

## Digitize new features

- In your file geodatabase, create a new point feature class named **PhippsTrees** whose spatial reference system is NAD_1983_StatePlane_Pennsylvania_South_FIPS_3702_Feet. In the PhippsTrees attribute table, create a text field named **Type**.
- In ArcMap, add the PhippsTrees features and the raster image from your file geodatabase. Using the ortho image as a background, digitize 15 or so trees in the front of Phipps Conservatory and assign them tree types Big Leaf Maple, Italian Cypress 1, and American Sycamore 2.

## Requirements

- In ArcScene, add all the features in your file geodatabase. Create a new TIN from the Topo layer named **Assignment10-3YourName_tin** and assign it spatial reference NAD_1983_StatePlane_Pennsylvania_South_FIPS_3702_Feet.
- Remove the original Topo layer and display the TIN as "Faces with the same symbol" (Early Desert color) and no edges or elevation.
- Drape the Bldgs, PhippsAddition, and PhippsTrees to the TIN using a layer offset of 0. Drape the ortho image and Curbs using a layer offset of 5. Depending on your computer's configuration, it might take some time to drape the ortho image.
- Extrude Bldgs and PhippsAddition polygon features using the height fields in each attribute table. The values in these fields are the building heights as opposed to the number of stories.
- Show the PhippsAddition using a transparency effect of 50% and the ortho image using a 25% transparency effect.
- Display the PhippsTrees point features as unique symbols using the corresponding 3D symbols for the three tree types.
- Create an animation of the site that eventually zooms into the new Phipps addition. Save the animation as an ArcScene animation file named **Assignment10-3YourName**.
- If instructed to do so, export the animation as **Assignment10-3YourName.avi**.

# 11

# ArcGIS Spatial Analyst for Desktop

This chapter is an introduction to ArcGIS® Spatial Analyst™, an extension of ArcGIS Desktop software. With Spatial Analyst, you can use or create raster datasets to display and analyze data that is distributed continuously over space as a surface. In this chapter, after an introduction to raster map layers, you'll prepare and analyze a demand surface map for the location of heart defibrillators in the City of Pittsburgh based on the number of out-of-hospital cardiac arrests. You'll also learn how to use ArcGIS Spatial Analyst to perform queries for a site suitability analysis and to create a poverty index surface that combines several census data measures from block and block group polygon layers.

Learning objectives

- Learn about raster map layers
- Create a hillshade map
- Make kernel density maps
- Extract point estimates from raster maps
- Conduct a raster-based site suitability study

## TUTORIAL 11-1
# Processing raster map layers

The map document that you will open has map layers that include raster maps from the US Geological Survey (USGS) for digital elevation using the National Elevation Dataset (NED shaded relief, 1/3 arc second) and land use from the 2006 National Land Cover Dataset. All raster maps are rectangular in their coordinate systems, but you will use Pittsburgh's boundary as a mask so that cells outside its boundary have no color and thus are invisible, and the cells inside have their assigned colors. In addition, you will use the NED layer to create a hillshade layer, which provides a 3D appearance of topography illuminated by an artificial sun. Placing the hillshade under the land-use layer and giving the land-use layer some transparency makes an attractive and informative display.

## Examine raster map layer properties

**1**  **Open Tutorial11-1.mxd from the Maps folder.** The vector map layer called OHCA (out-of-hospital cardiac arrests) is the number of heart attacks over a five-year period per census block that occurred outside hospitals where bystander help was possible because of location.

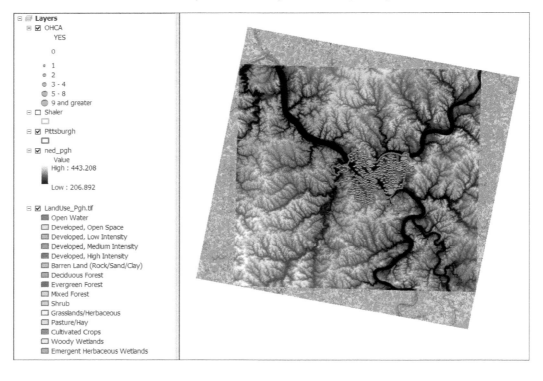

**2**   Save the map document to the Chapter11 folder of MyExercises.

Next, you'll examine properties of the raster layers.

**3**   In the table of contents, right-click ned_pgh and click Properties > Source tab. You can see in the Columns and Rows property that this raster map has 2,106 columns and 1,984 rows (and thus over four million cells displayed). You also can see in the Cell Size property that cells are square, 90.7 feet on a side. This layer is not so attractive or informative yet, but it will be after you transform it into a hillshade layer in the next tutorial of this chapter.

**4**   Scroll down until you see the Statistics properties. Each cell or pixel has a single value—elevation above mean sea level, feet in this case—which is stored as a floating point number. The statistics for elevation over the extent includes a mean of 323.7 feet, a minimum of 206.9 feet, and a maximum of 443.2 feet.

**5**   Cancel the Properties window.

## YOUR TURN

Examine the properties of the LandUse_Pgh layer. Notice in the Spatial Reference properties that this is a projected layer that uses a projection for the continental United States (this is why the layer tilts when ArcGIS projects it to the local state plane projection). Also, notice in the Raster Information properties that the cell size is about the same as the NED, 30 meters or 98.4 feet on a side, and that the values are integers corresponding to land-use categories.

## Set the raster environment

Next, you'll need to set the environment for using the ArcGIS Spatial Analyst tools. Each time you use one of the tools, ArcMap automatically applies the environmental settings, thereby saving you time and effort. First, you must turn on the Spatial Analyst extension, which is what you'll do in step 1.

**1**   Click Customize > Extensions > Spatial Analyst on > Close.

**2**   On the Menu bar, click Geoprocessing > Environments > Raster Analysis (you may have to scroll down in the Environment Settings window), and type or make selections as shown in the image, with As Specified Below for Cell Size and Pittsburgh for Mask.

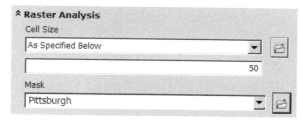

**3**   Click OK. Now Spatial Analyst automatically uses Pittsburgh as a mask and builds raster layers with cells 50 feet on a side. The unit is feet in this case because the data frame's projection is state plane in feet.

## Extract land use using a mask

ArcGIS Desktop can display many raster or image file formats. The land-use layer in your map document is a TIFF image downloaded from the USGS website. To process the layer, it must be converted into an Esri format, which you'll do by extracting the rectangular portion that is Pittsburgh's extent and saving that portion to a file geodatabase. At the same time, you'll use the Pittsburgh boundary as a mask to display cells only within Pittsburgh's boundary.

**1**   On the Menu bar, click Windows > Search. Type extract in the search text box, press Enter, and click Extract by Mask (Spatial Analyst) to open that tool.

**2**   Type or make selections as shown in the image, with LandUse_Pgh.tif for Input raster, Pittsburgh for Input raster or feature mask data, and LandUsePittsburgh for Output raster. This is one of the few tools that makes you select a mask, even though you set the mask in the environmental settings.

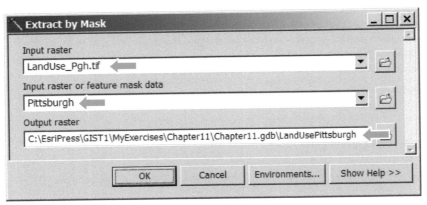

**3**   Click OK. Close the Search window.

**4**   In the table of contents, turn off all layers except OHCA and LandUsePittsburgh.

**5**   Right-click LandUsePittsburgh and click Zoom To Layer. ArcMap gave LandUsePittsburgh an arbitrary color ramp, but next you'll add a layer file to correctly symbolize the new raster map.

**6**  **Right-click LandUsePittsburgh, click Properties > Symbology tab > Import button** ![icon]**,
browse to the Data > SpatialAnalyst folder, click LandUse.lyr > Add, and click OK > OK.** The
resulting map is informative and attractive; for example, you can see high-density
development along Pittsburgh's rivers, and you can see that the clusters of heart attack
locations are in developed areas. In the next tutorial, you'll make the map even better by
giving it a 3D appearance, using hillshade based on the NED layer.

**7**   Save your map document.

---

## YOUR TURN

Extract LandUseShaler from LandUse_Pgh using the Shaler layer as a mask and saving it as LandUseShaler in \EsriPress\GIST1\MyExercises\Chapter11\Chapter11.gdb\. Symbolize the new layer using LandUse.lyr. When finished, turn off all the Shaler layers and zoom back to the Pittsburgh layer. *Hint:* You need to click Geoprocessing > Environments > Raster Analysis and change the mask to Shaler to do this work. The Extract tool does not work unless you make this change, even though you select Shaler as the mask in the tool. After the extraction is finished, be sure to set the Raster Analysis mask back to Pittsburgh. Save your map document and turn all Shaler map layers off in the table of contents.

## TUTORIAL 11-2
# Creating a hillshade raster layer

The hillshade function simulates illumination of a surface from an artificial light source representing the sun. Two parameters of this function are the altitude of the light source above the surface's horizon in degrees and its angle (azimuth) relative to true north. The effect of hillshade to a surface, such as elevation above sea level, is striking, giving a 3D appearance because of light and shadow. You can enhance the display of another raster layer, such as land use, by making land use partially transparent and placing hillshade beneath it.

## Create a hillshade for elevation

You will use the default values of the Hillshade tool for azimuth and altitude. The sun for your map will be in the west (315°) at an elevation of 45° above the horizon.

**1**  **Save your map document as** Tutorial11-2.mxd.

**2**  **Type** hillshade **in the search text box, press Enter, click Hillshade (Spatial Analyst), and close the Search window.**

**3**  **Type or make selections as shown in the image, with ned_pgh for Input raster and** HillshadePittsburgh **for Output raster.**

4   Click OK. Notice that Spatial Analyst used the Pittsburgh mask that you set in the raster environment. If your output does not display for Pittsburgh, check Geoprocessing, Environment Settings, and Raster Analysis to see if Pittsburgh is selected as the mask, and rerun Hillshade.

5   Right-click HillshadePittsburgh and click Properties > Symbology > Classified (in the Show panel) > Classify button (in the main panel).

6   Select Standard Deviation for Classification Method and 1/3 Std Dev for interval size. Click OK > OK.

7   In the table of contents, right-click LandUsePittsburgh and click Properties and the Display tab.

8   Type 35 in the Transparency field and click OK.

**9** Turn off the OHCA layer. Move HillshadePittsburgh to just below LandUsePittsburgh in the table of contents. That's the finished product. Hillshade provides a much richer-looking map.

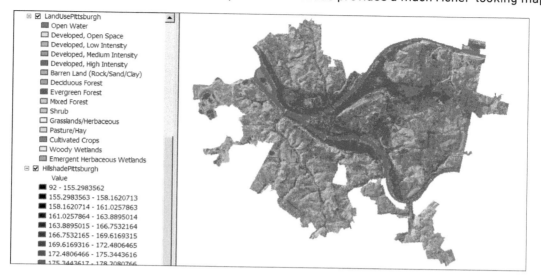

**10** Save your map document.

---

**YOUR TURN**

Create HillshadeShaler and display it under a 35 percent transparent LandUseShaler. Be sure to change the mask to Shaler in Geoprocessing > Environments and then back to Pittsburgh.

---

## TUTORIAL 11-3
# Making a kernel density map

The incidence of myocardial infarction (heart attacks) outside hospitals in the United States for ages 35 to 74 is approximately 5.6 per thousand males per year and 4.2 per thousand females per year.[1] You will use a point feature class of census block centroids in Allegheny County as input to an estimation method called *kernel density smoothing* to analyze heart attack incidence. This method estimates incidence as a density surface (heart attacks per unit area) and has two parameters, cell size and search radius.

# Make environmental settings and get statistics

**Note:** See the accompanying map before reading the paragraph at the start of this section.

The map document that you open in this tutorial shows actual and estimated values. The observed locations of heart attacks (outside of hospitals and thus with the potential of bystander assistance) are shown with size-graduated point markers at block centroids. Estimated heart attack incidence at block centroids is shown with a color gradient. The corresponding table of block centroids has the mapped incidence attribute, Inc = 0.0042 × [Fem35T74] + 0.0056 × [Male35T74], where Fem35T74 is the population by block of females ages 34 to 74, and Male35T74 is the population of males. The question is whether estimated incidence computed from residential locations does a good job of estimating the observed heart attacks in the OHCA point file. You will see right away that there are many locations in which estimated incidence is high but there are no observed heart attacks, so maybe the estimates aren't so accurate.

**1** **In ArcMap, open Tutorial11-3.mxd from the Maps folder.** The vector-based map display for estimated incidence using block centroids with point markers is as good as vector graphics allow. Nevertheless, the map is difficult to interpret; there are too many discrete points.

Next, you'll create an alternative representation of incidence by estimating a smoothed mean surface using kernel density smoothing. The result is a continuous map that is much easier to interpret and study than the vector map.

**2** **Save your map document as** Tutorial11-3.mxd **to the Chapter11 folder.**

**3** **On the Menu bar, click Geoprocessing > Environments > Raster Analysis, and type or make selections as shown , with As Specified Below for Cell Size and Pittsburgh for Mask.**

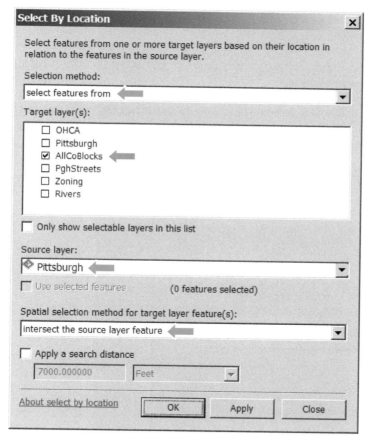

**☆ Raster Analysis**

Cell Size

| As Specified Below | ▼ | 📁 |

| 50 |

Mask

| Pittsburgh | ▼ | 📁 |

4   On the Menu bar, click **Selection > Select By Location >** and make selections as shown, with select features from for Selection method, AllCoBlocks **for Target layer(s), Pittsburgh for Source layer, and** intersect the source layer feature for Spatial selection method for target **layer(s) features.**

**Select By Location**                                                      ✕

Select features from one or more target layers based on their location in relation to the features in the source layer.

Selection method:

| select features from  ⬅ | ▼ |

Target layer(s):

☐ OHCA
☐ Pittsburgh
☑ AllCoBlocks  ⬅
☐ PghStreets
☐ Zoning
☐ Rivers

☐ Only show selectable layers in this list

Source layer:

| ◇ Pittsburgh  ⬅ | ▼ |

☐ Use selected features          (0 features selected)

Spatial selection method for target layer feature(s):

| intersect the source layer feature  ⬅ | ▼ |

☐ Apply a search distance

| 7000.000000 | Feet | ▼ |

About select by location        [ OK ]    [ Apply ]    [ Close ]

5   Click OK.

6   In the table of contents, right-click AllCoBlocks and click Open Attribute Table, and then right-click the header for Inc and click Statistics. Note the sum of Inc, 684 for Pittsburgh, which is the expected annual number of heart attacks in Pittsburgh outside hospitals in a year's time.

In the next tutorial, you'll verify that density smoothing preserves this sum in any surface it estimates. Kernel density smoothing simply spreads the total around on a smooth surface, preserving the input total number of heart attacks.

**7**  **Close the Selection Statistics window and the table and clear the selection.**

## Make a density map for heart attack incidence

The OHCA map layer shows heart attacks per census block in Pittsburgh. Blocks in Pittsburgh average a little less than 300 feet per side in length. Suppose that policy analysts estimate that a defibrillator with public access can be made known to residents and retrieved for use as far away as 2.5 blocks from the location. They thus recommend looking at areas that are five blocks by five blocks in size, or 1,500 feet on a side, with defibrillators located in the center. Therefore, you can use a 150-foot cell and a 1,500-foot search radius. The 150-foot cell approximates the middle of a street segment, the average location of a heart attack.

**1**  **Type** kernel density **in the search text box, press Enter, click Kernel Density (Spatial Analyst), and close the Search window.**

**2**  **Type or make selections as shown in the image, with AllCoBlocks for Input point or polyline features, Inc for Population field,** KernelDensity1500 **for Output raster,** 150 **for Output cell size, DENSITIES for Output values are, and PLANAR for method.**

**3** **Click OK.** The resulting surface will appear useful after you symbolize it better in the following step.

**4** **Right-click KernelDensity1500 and click Properties > the Symbology tab > the Classify button. Select Standard Deviation for Classification Method.** In general, standard deviation is a good option for showing variation in raster grids because it yields a central category and categories on each side of the center. That makes dichromatic color scales more meaningful and easier to interpret. You control the number of categories in the next step by choosing the fraction of standard deviation for creating break points, every 1, 1/2, 1/3, or 1/4 standard deviation.

Next, you'll use dichromatic color scales.

**5** **Select 1/3 Std Dev for Interval Size and click OK.**

**6** **Select the color ramp that runs from green to yellow to red and click OK.**

**7** **Turn off AllCoBlocks.** Incidence matches clusters of the OHCA heart attack data in many but not all areas. For example, there is a cluster in Pittsburgh's Central Business District (triangle just to the right of where the three rivers meet), but estimated incidence is low there. The problem is that the density map, based on population data, shows expected heart attacks per square foot in reference to where people live, not necessarily where they have heart attacks. Many people shop or work in the Central Business District and unfortunately have heart attacks there, but few live there.

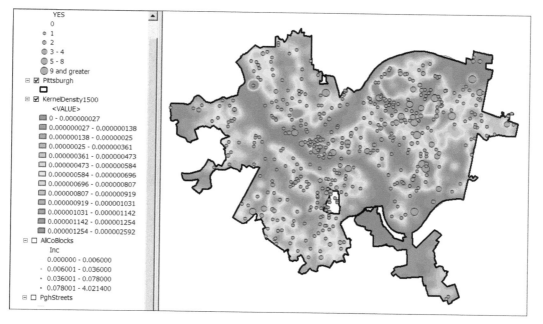

Check the density surface to see if it preserves the total number of heart attacks. Note that your values may vary slightly. The estimated incidence that you found using block centroids was 684. Open the properties for the density surface and click the Source tab. There you will find the mean of 0.000000420184 heart attacks per square foot. To get the number of cells in the Pittsburgh mask, using a shortcut will help. Select Symbology in the Layer Properties window and click Classify. There you will find that there are 72,315 cells in the Pittsburgh mask. (The estimate of the mean density in this display, 0.000000417, is less accurate than the one from the Source tab.) Remember that each cell is 150 feet by 150 feet. Therefore, 150 × 150 × 72,315 × 0.000000420184 = 684 smoothed heart attacks. So kernel density smoothing moved around the input number of heart attacks located at block centroids to distribute them smoothly. The kernel density map is a better estimate of incidence than raw data, because smoothing averages out spatial randomness and provides an estimate of the mean or average surface. Close the Classification and Layer Properties windows.

## YOUR TURN

Create a second kernel density surface for incidence, named KernelDensity3000, with all inputs the same except you'll use a search radius of 3,000 instead of 1,500. Symbolize the output the same as Density1500. While keeping KernelDensity1500 turned on, turn KernelDensity3000 on and off to see the differences in the two layers. KernelDensity3000 is more spread out and smoother, but it has the same corresponding number of estimated heart attacks: close to 684. Save your map document.

## TUTORIAL 11-4
# Extracting raster value points

Although the estimated densities appear to match the actual heart attack data in OHCA, the match may or may not stand up to closer investigation. ArcMap has a tool that extracts point estimates from the raster surface for each point in OHCA. Then you can use the extracted densities multiplied by block areas to estimate the number of heart attacks. If there is a strong correlation between the estimated and actual heart attacks, evidence would suggest that population alone is a good predictor of heart attacks outside hospitals.

**1**  **Save your map document as** Tutorial11-4.mxd **to the Chapter11 folder.**

**2**  **Type** extract values to points **in the search text box, press Enter, click Extract Values to Points (Spatial Analyst), and close the Search window.**

**3**  **Type or make selections as shown in the image, with OHCA for Input point features, KernelDensity1500 for Input raster,** OHCAPredicted **for Output point features, and "Interpolate values at the point locations (optional)" checked .**

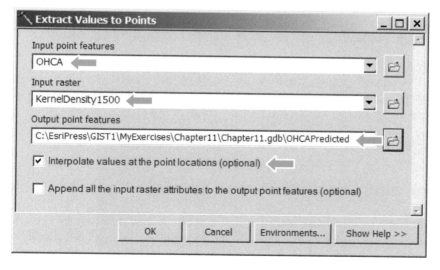

**4**  **Click OK.** The resulting layer, OHCAPredicted, has an attribute, RASTERVALU, which is an estimate of heart attack density, or heart attacks per square foot, in the vicinity of each block.

## Calculate predicted heart attacks

**Note:** Adding more significant digits to RASTERVALU, as you'll do in the revised step, helps you understand the rest of the steps in this tutorial in which you multiply RASTERVALU by two other quantities. As originally displayed, it looks as if there are rows in which RASTERVALU = 0, but the result of multiplication by other nonzero values is nonzero (impossible!). However, when you display more significant digits, you'll see that all RASTERVALU values are positive (none are zero).

**1** **Right-click OHCAPredicted and open its attribute table. Right-click the RASTERVALU column heading, click Properties, click the numeric builder button ⌞…⌟ , and select the Number of significant digits option button. Select 9 for number of decimal places, and click OK > OK.**

**2** **Click Options, Add Field, and add a field named** Predicted **with the Float data type.**

**3** **Right-click the Predicted header and click Field Calculator.**

**4** **Create the expression** 5 * [RASTERVALU] * [Area] **and click OK.** OHCA data is a five-year sample for heart attacks, so the expression includes the multiple 5. Note: [Area] is the area of the city block being represented by a point in OHCA. So, the density, [RASTERVALU] (heart attacks per square foot), times [Area] (square feet) yields the number of heart attacks.

**5** **Close the attribute table.** A few of the points in OHCA have no raster values near them, so ArcGIS assigns −9999 to their RASTERVALU cells to signify missing values. Before looking at a scatter plot of predicted and actual values, you'll first select only OHCA points with positive predicted values.

**6** **Click Selection > Select By Attributes.**

**7** **For the OHCAPredicted layer, create the expression** Predicted >= 0 **and click OK.**

**8** **Right-click OHCAPredicted and click Data > Export Data.**

**9** **Export selected features to \EsriPress\GIST1\MyExercises\Chapter11\Chapter11.gdb\** OHCAPredicted2 **and click Yes to add the feature class to the map.**

**10** **Clear the selected features and turn off the OHCAPredicted layer.**

## Create a scatter plot of actual versus predicted heart attacks

Although Pittsburgh has a total of 7,466 blocks, only 1,509 blocks had heart attacks. The scatter plot that you eventually will construct includes data for only the 1,509 blocks but should include the balance of the total blocks, which had actual values of zero but predicted values sometimes much larger than zero. Nevertheless, you can get an indication of the correlation between predicted and actual heart attacks. Adding the balance of blocks would only make the correlation worse, but the correlation is already low, as you'll see next.

**1** **Click View > Graphs > Create Graph.**

**2** **Type or make selections as shown, with Scatter Plot for Graph type, OHCAPredicted2 for Layer/Table, Yes for Y field, Predicted for X field (optional), and Add to legend box unchecked.**

3   **Click Next > Finish.** At the scale of blocks, the predicted values correlate poorly with the actual values. A good correlation would have a graph with Actual (YES attribute) and predicted values scattering around a 45-degree slope line. This scatter plot shows no correlation at all. If you export the corresponding data to a statistical package or Excel, you would find that the correlation coefficient between predicted and actual values is only 0.0899, which is low. Evidently, factors other than where the population resides affect the locations and clustering of heart attacks occurring outside hospitals.

4   **Close the graph and save your map document.**

## TUTORIAL 11-5
# Conducting a raster-based site suitability study

The objective is to find locations that have high heart attack rates and that are candidates to have heart defibrillators accessible to the public. The approach starts by using kernel density smoothing on the available heart attack (OHCA) data to remove spatial randomness from the spatial distribution and provide a more reliable estimate of demand. An assumption is that any locations within commercial areas provide public accessibility. Efforts then turn to querying the kernel density map of heart attacks for areas with sufficiently high densities that are in commercial areas. To query raster maps, you must reclassify cells to 1 and 0 values (representing true and false) and use Boolean operators, such as AND and OR, to combine raster maps and meet query criteria.

## Open a map document

A vector map layer is available for commercial area boundaries. To conduct a raster-based analysis, you must convert this map layer into a raster layer.

**1    In ArcMap, open Tutorial11-5.mxd from the Maps folder.** The map document shows the observed locations of heart attacks outside hospitals, a 600-foot buffer of commercially zoned areas in Pittsburgh, and other supporting layers. The 600-foot (or two-block) buffer of commercial areas includes adjacent noncommercial areas that have adequate access to defibrillators.

**2**   Save the map document to the Chapter11 MyExercises folder.

**3**   On the Menu bar, click Geoprocessing > Environments > Raster Analysis and type or make selections as shown in the image, with **As Specified Below** for Cell Size and **Pittsburgh** for Mask.

**4**   Click OK.

# Convert a feature buffer layer to a raster dataset

The ZoningCommercialBuffer layer has two polygons and corresponding records with a single attribute: Commercial. The Commercial value of 1 corresponds to commercial land use or land within 600 feet of commercial land use. The other value, 0, represents the balance of Pittsburgh and includes all other zoned land uses. You will convert this vector layer into a raster dataset using a conversion tool. First, however, you need to select both records in the vector file for them to convert.

**1**   Right-click the ZoningCommercialBuffer layer and click Open Attribute table.

**2**   Select both records by clicking the row selector of the first row and dragging the mouse to select both rows, and then close the table.

**3**   Type polygon to raster in the search text box, press Enter, click Polygon to Raster (Conversion), and close the Search window.

**4** Type or make selections as shown in the image, with **ZoningCommercialBuffer** for **Input Features**, **Commercial** for **Value field**, CommercialRaster **for Output Raster Dataset, CELL_ CENTR for Cell assignment type, NONE for Priority field, and 50 for Cellsize.**

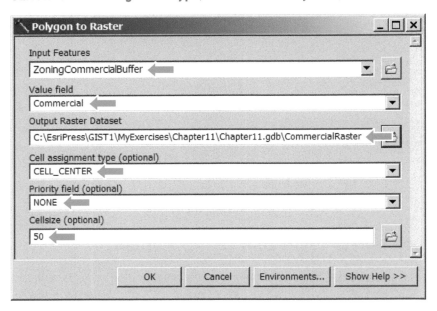

**5** Click OK.

**6** Remove the ZoningCommercialBuffer layer and turn off the OHCA layer.

**7** Right-click CommercialRaster, click Properties > the Symbology tab > Unique Values in the Show panel, and resymbolize the new Commercial area with two colors: white for uncommercial and gray for commercial.

## YOUR TURN

Create a kernel density map based on the YES attribute of the OHCA point layer that has 150-foot cells, a search radius of 1,500 feet, and area units of SQUARE_FEET. Call the new raster layer HeartAttack and save it to Chapter11.gdb. Symbolize the layer using the standard deviation method with interval size 1/3 Std Dev. Use the green-to-yellow-to-red color ramp. Try turning the OHCA layer on and off to see how well the density surface represents heart attacks, and then remove the OHCA layer.

# Query a raster dataset with a single criterion using reclassify

First, you'll reclassify your kernel density map, HeartAttack, for areas that have high enough heart attack density to merit a defibrillator. Suppose that policy makers seek 25-block areas, roughly five blocks on a side, that would have 10 or more heart attacks every five years in locations in which bystander help is possible. A square 25-block area is 5 × 300 feet = 1,500 feet on a side with 1,500 feet × 1,500 feet = 2.25 × 106 square feet of area.

Thus, the heart attack density sought is 10 heart attacks / 2.25 × 106 square feet = 0.000004444 heart attacks per square foot or higher. Although the density map you just created has a continuous range of values, next you'll reclassify values into just two values: 0 for cells with density less than 0.000004444 and 1 for cells with density greater than or equal to 0.000004444.

**1** Type reclassify in the search text box, press Enter, click Reclassify (Spatial Analyst), and close the Search window.

**2** Select HeartAttack for the input raster and click Classify.

**3** Select 2 for Classes, select Manual for Method, type 0.000004444 to replace 0.000006 (the first of two values) in the Break Values panel, and click OK. The Old values column shows values with only 6 decimal places, but the Reclassify tool has all 9 decimal places saved.

**4** Click Precision, select 9 for the number of decimal places, and click OK.

**5**   Finish filling in the form by typing or making selections as shown in the image, with
HeartAttack for **Input Raster**, 0 and 1 under **New Values**, and HAttackQ1 **for Output raster** .

**6**   Click OK.

**7**   Resymbolize HAttackQ1 so that 0 has no color and 1 is dark blue, and make sure that
HeartAttack is turned on and below HAttackQ1. In the image, you can see that relatively few
peak areas (6) have high enough heart attack density and may be large enough in area.

## Query a raster dataset with two criteria

Next, you'll include a second criterion in the query—locations within the commercial buffer—for suitable defibrillator sites. Recall that the CommercialRaster map has 0 for noncommercial areas and 1 for commercial areas. In the previous section, you just set dense enough areas for heart attacks to 1 and those not dense enough to 0. The Boolean AND tool combines two raster datasets by giving all cells the value 0 except where both input cells are 1, in which case the output cell gets the value 1. In this case, the resulting areas defined by cells with value 1 are both in the commercial buffer and have the high enough heart attack density of HAttackQ1.

1   Search for Boolean and in the search text box. Notice that all the Boolean logical operators are available as tools, including NOT, AND, and OR.

2   Click Boolean And (Spatial Analyst), and close the Search window.

**3** Type or make selections as shown in the image, with **HAttackQ1** for **Input raster or constant value 1**, **CommericalRaster** for **Input raster or constant value 2**, and HAttackQ2 **for Output raster.**

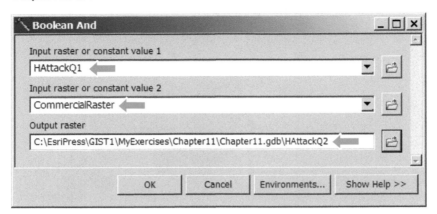

**4** Click OK.

**5** Resymbolize HAttackQ2 so that 0 has no color and 1 is Tourmaline Green (column 8, row 3 of the color chips array), and make sure that HAttackQ1 and HeartAttack are turned on and below HAttackQ2. As you would expect, adding a second criterion with the AND connection has reduced the size of areas meeting criteria. Three of the formerly promising areas are significantly reduced.

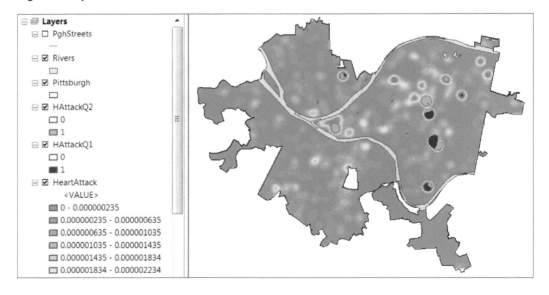

**6** Save your map document.

## YOUR TURN

Turn on the Streets layer and zoom in to each feasible area to check the third criterion that there be at least 25 blocks, or roughly 2.25 million square feet, in a square area. Use the Measure tool on the toolbar to measure feasible areas. Which areas remain feasible? What would you report back to policy makers? Save the map document.

# Create a mask and hillshade for suburbs

The first ring of suburbs around urban areas is a good area for revitalization as aging suburban homeowners attempt to downsize houses and relocate closer to work. The houses in these areas tend to be relatively small and well-constructed but in need of renovation.

This assignment has you choose a subset of municipalities in Allegheny County comprising the first ring of suburbs around Pittsburgh to display land use with hillshade. You'll also display parks for the suburbs. The resulting map document and layers will provide a good starting point for redevelopment work.

## Get set up

- Create a new folder, **\EsriPress\GIST1\MyAssignments\Chapter11\Assignment11-1YourName\** and store all files that you produce for this assignment in it.
- Create a map document named **Assignment11-1YourName.mxd** with relative paths.
- Create a new file geodatabase named **Assignment11-1YourName.gdb** and store finished files that you create in this assignment in it. Do not include or hand in any of the raw data shapefiles in the next bulleted list.

## Build the map

Add the following to your map document. Add the municipalities first, so that your data frame inherits that layer's projection, which is the local 1983 state plane projection.

- \EsriPress\GIST1\Data\AlleghenyCounty.gdb\Munic—polygon features for municipalities in Allegheny County.
- \EsriPress\GIST1\Data\AlleghenyCounty.gdb\Parks—polygon features for parks. DESCR is an attribute with park name.
- \EsriPress\GIST1\Data\SpatialAnalyst\SpatialAnalyst.gdb\NED—digital elevation model for a portion of Allegheny County.

- \EsriPress\GIST1\Data\SpatialAnalyst\SpatialAnalyst.gdb\Pittsburgh—polygon feature for Pittsburgh's municipal boundary.
- \EsriPress\GIST1\Data\SpatialAnalyst\LandUse\LandUse_Pgh.tif—land use for a portion of Allegheny County.

## Requirements

- Turn off all layers except Munic to simplify the next step, in which you create a ring of suburbs. Define the first ring of suburbs as those within 1.25 miles of Pittsburgh, but not including Pittsburgh. Start by making Munic the only selectable layer. Then use Selection > Select By Location, and select municipalities that are within a distance of 1.25 miles of Pittsburgh. Use the Select Features tool, press and hold your Shift key, and click inside the Pittsburgh polygon in the Munic layer to deselect it. You won't see any change in the selection because surrounding municipalities share Pittsburgh's boundary, but the Pittsburgh polygon should no longer be selected. Also deselect Mount Oliver, which is the polygon inside Pittsburgh and not a suburb. Finally, right-click Munic, click Data > Export data to create **Suburbs**, and add it to your map document. Now select Parks that intersect with suburbs and create **SuburbanParks**.
- Using Suburbs as the mask and a cell size of 50, extract a raster from LandUse named **LandUseSuburbs**. Create a hillshade from NED named **HillshadeSuburbs**.

The following are additional instructions on symbolization:

- To symbolize LandUseSuburbs, click Unique Values in the Show panel of the Layer Properties Symbology tab, click the Import button (file icon), browse to the Data > SpatialAnalyst folder, and import LandUse.lyr.
- Use hollow fill and a thin outline for Suburban Parks. Label them with their names with the smallest font size when zoomed in to 1:100,000 or closer (see the Labels tab of the properties sheet and its Scale Range button).
- Symbolize Suburbs with a hollow fill and thick outline. Label with a small font and a halo.
- Symbolize HillshadeSuburbs using Classified in the Symbology show panel with the Standard Deviation classification method and 1/3 Std Dev interval size.
- Make LandUseSuburbs transparent, move it above the hillshade, turn off unneeded layers, and display suburban parks with the shaded land-use layer. Housing will be in the red, developed areas.

# Estimate heart attack fatalities outside hospitals by gender

Unfortunately, females have more fatal heart attacks outside hospitals than males, perhaps because symptoms of heart attacks in females are less well known and more subtle than those for males. Heart attacks outside hospitals are roughly 1.5 per thousand for males ages 35 to 74 and 2.3 per thousand for females in the same age range. In this assignment, you create two density map layers—one for males and one for females—using these incidence rates for the municipality of Wilkinsburg in Allegheny County. You do all raster processing using Spatial Analysis tools in a model.

## Get set up

- Create a new folder, **\EsriPress\GIST1\MyAssignments\Chapter11\Assignment11-2YourName\** and store all files that you produce for this assignment in it.
- Create a map document named **Assignment11-2YourName.mxd** with relative paths.
- Create a new file geodatabase named **Assignment11-2YourName.gdb** and store files that you create in this assignment in it. Do not include or hand in any of the raw data shapefiles in the next bulleted list.

## Build the map

Add the following to your map document:

- \EsriPress\GIST1\Data\AlleghenyCounty.gdb\Munic—polygon features for municipalities in Allegheny County.
- \EsriPress\GIST1\Data\SpatialAnalyst\SpatialAnalyst.gdb\AllCoBlocks—point features for census block centroids in Allegheny County.

## Requirements

- Select the Wilkinsburg polygon from Munic and export it as **Wilkinsburg**. Extract the Wilkinsburg blocks from AllCoBlocks and save them as **WilkinsburgBlocks**. Add floating point fields to the attribute table for WilkinsburgBlocks: **MMortinc = 0.0015 × [Male35T74]** for the annual number of heart attack fatalities for males ages 35 to 74 and **FMortinc = 0.0023 × [Fem35T74]** for females ages 35 to 74.
- Create kernel density map layers for MMortinc and FMortinc using Wilkinsburg as the mask and with a cell size of 25 and search radius of 1,500 feet. Add outputs to the map.

## Additional instructions

- After you extract Wilkinsburg, to give your data frame Wilkinsburg as its full extent instead of Allegheny County, do the following: right-click Layers in the table of contents and click Properties > the Data Frame tab. Select Fixed Extent for Extent, click Specify Extent, the Outline of Features option button, select Wilkinsburg as the layer, and click OK > OK.
- Set the environment for ArcToolbox. For Workspace, make \MyAssignments\Chapter11 \ Assignment11-2YourName\ the Current Workspace and set the extent to Wilkinsburg. For Raster Analysis, set the cell size to 25 and set the Mask to Wilkinsburg.
- Set Geoprocessing options. Make sure that "Overwrite the outputs of geoprocessing operations" and "Add results of geoprocessing operations to the display" are selected.
- Use ArcCatalog to create the new toolbox, **DiseaseIncidence.tbx**, and add a model to it. Then, in ArcToolbox, right-click the ArcToolbox icon at the top of the ArcToolbox window, click Add Toolbox, and add your new toolbox.
- Your ModelBuilder model is a simple one. It has only two processes, each a kernel density estimation producing the two output rasters. **Hint 1:** Raster layer names cannot be longer than 13 characters. **Hint 2:** Right-click each output raster in the model and click Add To Display so that you don't have to add the outputs yourself.
- Symbolize the female mortality surface using Classified from the left panel of the Symbology tab of Layer Properties. Use the Standard Deviation Classification method with 1/3 standard deviation intervals. Then, for male mortality, import symbolization from the female mortality layer so that both layers have identical break points and the same color ramp. Then you easily can compare female and male incidence by turning one of them off and on while the other is on.
- **Hint:** If you need to rerun your model, click Model > Validate Entire Model.

### Note

1. Wayne D. Rosamond et al., "Trends in Incidence of Myocardial Infarction and in Mortality Due to Coronary Heart Disease, 1987 to 1994," *New England Journal of Medicine* 339, no. 13 (1998): 861–67.

# 12

# ArcGIS Network Analyst for Desktop

The ArcGIS® Network Analyst™ extension of ArcGIS Desktop software implements network optimization models from the applied mathematics field of operations research, including: (a) the traveling salesperson problem, in which a salesperson needs the shortest route for making several stops before returning to home base; (b) the facility location problem, for choosing one or more locations for new facilities while taking into account demographics and the possibility of existing facilities including those of competitors; and (c) the multivehicle routing problem, with one or more warehouses/depots and many stops to meet the needs of a set of stores or other facilities. Although network optimization models and algorithms were developed nearly 50 years ago, they were not useful until the advent of GIS and map infrastructure, including street networks, were made available. Network Analyst is one of the few practical implementations of these powerful models coupled with GIS.

Network optimization models depend on network datasets, which are based on street layers (and other transportation layers such as metro transit lines), to include data on connectivity of line segments, restrictions to flow along line segments such as one-way streets, data on average or real-time travel speeds on network segments, and other data such as allowable turns at intersecting line segments.

Accurate and up-to-date street network data is available commercially but is too expensive for individuals to purchase or for Esri to package with ArcGIS. Nevertheless, there are sources of network data for educational purposes and projects that do not depend on either having the latest or most accurate data. One option is that you can build your own street network dataset for a city or county, based on TIGER street centerlines (as done in tutorial 12-2). Its limitations are that it does not include data on one-way streets and turn restrictions at intersections, and it uses only typical travel speeds for street or road types instead of estimates of actual travel speeds. It also uses only typical time delays for moving through or turning at intersections. A second option is that Esri provides sample street network datasets for educational purposes, including one for San Francisco used in tutorials 12-1 and 12-5 of this chapter.

### Learning objectives

- Learn about network optimization models
- Formulate and solve the traveling salesperson problem
- Build a network dataset
- Create service area/travel polygons
- Locate facilities
- Route multiple delivery vehicles

## TUTORIAL 12-1
# Solving the "traveling salesperson" problem

The optimization problem of this tutorial is stated in terms of a traveling salesperson but applies to many routing problems. The salesperson has a home base and several stops to make during a run before returning to home base. Needed is the best sequence and routes to minimize travel time. Although there is no exact solution for this combinatorial problem, there are good heuristic (approximate) algorithms, and one of the best is implemented in Network Analyst. (Similarly, the solution methods for the other models of Network Analyst also use heuristic algorithms and provide very good if not optimal solutions.)

In the following exercises, we consider the hypothetical case of a commercial kitchen, Kate's Wholesome Meals Kitchen in San Francisco, which prepares meals for home delivery ordered by phone or via the internet. Two or more times a day, the kitchen sends out a truck to deliver meals to a collection of addresses. The workflow to solve this problem in Network Analyst has several steps that are standard for all Network Analyst models: (1) configure the network and load location data, (2) choose a network model, (3) choose property (parameter) values, and (4) solve the problem.

# Get started

**1  Open Tutorial 12-1.mxd from the Maps folder.** The map document has a network dataset named Streets_ND for San Francisco. which includes historic average travel times by weekday versus weekend as well as restricted flow directions along one-way streets. Kate's Kitchen has the large point marker with a tablespoon and knife. Run21097 is a point layer geocoded by street address for the 12 meal orders to be delivered in a particular run. The run's orders have ID numbers 1741 through 1752, as seen in the layer's labels, which also include street addresses. Needed is the best sequence and routes to visit all 12 stops and then return to the kitchen.

**2  Save the map document to the Chapter12 folder of MyExercises.** Next, you'll have to enable the Network Analyst extension in ArcMap and open its toolbar.

**3  Click Customize > Extensions > Network Analyst > Close.**

**4  Click Customize > Toolbars > Network Analyst.** You can leave the toolbar as floating or dock it by dragging its gray top to a location.

# Choose a model and load stops

Stops are geocoded from street address data, using the methods of chapter 7, and then added to the network. In real-world applications, an information system would save customers' addresses and coordinates for reuse without needing new address matching. When a new customer is added, or changes his or her delivery address, the address would be geocoded with coordinates, and then added to the master customer table. Suppose that customers already have been selected from the master list to make the layer for run 21097.

**1** Click Network Analyst on the left side of the toolbar and select New Route. Network Analyst creates a Route group layer in the table of contents that will display the solution and related information after the model is run.

**2** Click the Network Analyst Window button ⊞ on the toolbar, right-click Stops (0), and click Load Locations. The Load Locations window opens where you will load the points of Run21097 as the stops for the route.

**3** Select Run21097 for the Load From drop-down menu and click OK. Expand "Stops" in the Network Analyst window to see the 12 delivery stops, plus two kitchen stops at 1117 Howard St. You'll need the kitchen to be the first and last stop, so next, you will move the second 1117 Howard St stop to the bottom of the list.

**4** Click the second 1117 Howard St stop and drag it to the bottom of the Stops list. In the next exercise, you'll instruct Network Analyst to preserve the first and last stops, while finding the minimum-time route through all 12 delivery stops. That makes this a traveling salesperson problem.

## Set model parameters

**1** Click the Route Properties button ⊞ in the upper-right corner of the Network Analyst window.

**2** Click the Analyst Settings tab and select or type settings as shown, with WeekdayFallbackTravelTime for Impedance, the Reorder Stops to Find Optimal Route box checked, and the Use Hierarchy box unchecked.

**3** **Click OK.** The critical setting for the traveling salesperson problem is to reorder stops while preserving the first and last stops. The reorder stops setting results in optimal stops.

## Solve the problem

With data made available and parameters set, you are ready to run the model. As you'll see, you easily can run the model over and over again, making various changes, and immediately see the results—which is helpful.

**1**    **Click the Solve button ⊞ on the Network Analyst toolbar.** Network Analyst finds an appealing-looking route. The challenge for the solver was finding a good route from stop 3 to stop 4. Apparently, the shortest route was through the Golden Gate National Recreational Area. Also, the route from stop 6 to 7 and then from stop 7 to 8 has the driver double back along part of the route instead of taking a direct route from 7 to 8. There is a park without streets in the way of a direct route.

**2**    **Click the Directions button ⬆ on the Network Analyst toolbar.** Shown in the figure is a partial listing of the resulting turn-by-turn directions available by printout for the driver. Included is the optional map for stop 7, 1663 Turk St. The total route length is 19.8 miles with exactly one hour of travel time. If it takes an average of five minutes per stop to park and deliver an order, then, all told, there's an additional one hour of work for a total of two hours for the run.

**3** Close the Directions window.

> ## YOUR TURN
> Rerun the model using weekend fallback travel times (instead of weekday fallback times) in the Analysis Settings. After making the change, press the Solve button and open the Directions window. In this case, the same route is optimal, but it takes two minutes (only) less time to traverse because weekend traffic is less crowded than weekday traffic. When finished, reset to weekday fallback times and rerun the model.

## Add barrier

Suppose that the manager at Kate's Kitchen learns that there's a dog-walk-for-charity event in Golden Gate National Recreational Area, and some of the park's streets will be blocked off. You'll add a barrier to the park's streets, and then re-solve the problem.

**1**   If not already open, open the Network Analyst window, scroll down, and select Restriction (0) under Point Barriers. Next you'll use the Create Network Location Tool and, considering that you've selected a point barrier restriction, when you click the map with the tool, it will add a barrier.

**2**   Zoom in to the location shown in the image preceding step 1. Turn off Routes in the table of contents so that you can see the streets below the route.

**3**   Click the Create Network Location button ⬝⛰. Click Windows > Magnifier, move the Magnifier window to the area as shown in the figure, release the Magnifier window so that it zooms in, and click the street on the left that runs north and south. That adds the graphic pick you just made as a point barrier restriction. If you make a mistake, click the Undo button and try again.

**4**  **Close the Magnifier window, turn Routes back on, and re-solve the problem by clicking the Solve button. Use the Run21097 bookmark to see the entire route.** The solver makes a change to avoid the road block, and if you open the Directions window, you'll see that the change reduces total travel distance from 19.8 miles to 19.0 miles but adds a minute to travel time, making it one hour and one minute. The new partial route must have slower travel times and/or more intersections.

**5**  Save your map document.

## YOUR TURN

See if you can find a location to include as a barrier in the current route that causes a large change. You can delete a trial barrier in the Network Analyst window by right-clicking its graphic pick under Point Barriers Restriction and clicking Delete. When finished, save your map document.

## TUTORIAL 12-2
# Building a TIGER-based network dataset

The GIST1 folder includes a TIGER street layer for San Francisco that you will use in this tutorial to build a network dataset. After you have the network dataset built, you'll rerun the orders for Kate's Wholesome Food Kitchen and see how the resulting route compares with the one you created in tutorial 12-1, which used a more sophisticated network dataset.

## Get started

1 **Open Tutorial 12-2.mxd from the Maps folder.** The map document has many of the same layers as in Tutorial 12-1, but instead of a network dataset, it has TIGER street centerlines (the SFStreets layer). Also new in the map document is the table MFTCC$ that has MFTCC codes (MAF/TIGER Feature Class Code) for streets and typical travel speeds in miles per hour. See the accompanying table. So rather than being calibrated with estimates for actual travel speeds, the network you will build uses the typical speeds from this table. This will produce some errors in solutions, but nevertheless will represent the street network of the city. Note the low speeds, 0.001 miles per hour, of pedestrian trails and stairways. Those values are included so that when routing vehicles, those features are not used (they are so slow that other street features would be chosen). If, however, one were calibrating the network for walking, then most MFTCC street types, including trails and stairways, could be given an average walking speed of 3.1 miles per hour.

| MTFCC | SPEED | DESCRIPTION |
|-------|-------|-------------|
| S1100 | 55 | Interstate highway or primary road with limited access |
| S1200 | 45 | Primary road without limited access |
| S1400 | 25 | Neighborhood road, rural road, city street |
| S1500 | 10 | Vehicular trail |
| S1630 | 15 | Ramp |
| S1640 | 15 | Service drive |
| S1710 | 0.001 | Pedestrian trail |
| S1720 | 0.001 | Stairway |
| S1730 | 5 | Alley |
| S1740 | 5 | Private road |

**2**  Save the map document to the Chapter12 folder of MyExercises.

**3**  If necessary, click Customize > Extensions, click Network Analyst on, and click Close. Also click Customize > Toolbars > Network Analyst.

## Add network fields to SFStreets

You'll add attributes to SFStreets for times in minutes to traverse each street segment in both directions. Street segments have "to" and "from" directions, and network datasets have the capacity to use different travel times for each direction. So whereas the network you'll build requires times for both directions, you'll use the same times for each direction based on the Speed attribute of the MFTCC$ code table. You must divide length by speed to get time:

Time (Minutes) = Length (Feet)/[Speed (Miles/Hour) × 5280 (Feet/Mile)/60 (Minutes/Hour)] = Length/[Speed × 88]

**1**  If the MTFCC$ sheet shows up with a red exclamation point, you will need to save the file as another file type. Use File Explorer to locate the file StreetSpeeds.xlsx in Data\ NetworkAnalyst and open it in Excel.

**2**  With the MTFCC sheet selected, save the file with the .csv file type. Click OK to save only the active sheet. Close Excel.

**3**  In Catalog, refresh the Network Analyst folder. Add the new .csv file to the table of contents and remove the old file.

**4**  Open the attribute table of SFStreets and create three new fields: Length, TF_Minutes, and FT_Minutes, all with the Float data type.

**5**  Right-click SFStreets in the table of contents, click Joins and Relates > Join, and make selections or type as shown in the following image, with MTFCC for no. 1, StreetSpeeds.csv for no. 2, and MTFCC for no. 3. Here you are joining the code table to streets. You'll use the Speed column from this table with street segment lengths that you'll calculate to compute travel times per street segment.

**12**

Join Data                                                    ✕

Join lets you append additional data to this layer's attribute table so you can,
for example, symbolize the layer's features using this data.

What do you want to join to this layer?

| Join attributes from a table                                          ∨ |

1. Choose the field in this layer that the join will be based on:

| MTFCC ◀━━━                                                            ∨ |

2. Choose the table to join to this layer, or load the table from disk:

| StreetSpeeds.csv ◀━━━                                    ▼ |  📂

☑ Show the attribute tables of layers in this list

3. Choose the field in the table to base the join on:

| MTFCC ◀━━━                                                            ∨ |

Join Options
◉ Keep all records
   All records in the target table are shown in the resulting table.
   Unmatched records will contain null values for all fields being
   appended into the target table from the join table.

○ Keep only matching records
   If a record in the target table doesn't have a match in the join
   table, that record is removed from the resulting target table.

|                          Validate Join                          |

About joining data               |        OK        |   |   Cancel   |

**6**   **Click OK.** Note: If you cannot scroll to the right in the SFStreets table in the next step, close
and then reopen the table.

**7**   **Right-click the Length field heading in the SFStreets attribute table, click Calculate
Geometry, make sure that Length is selected for Property and Feet US [ft] for Units, and
click OK.** That calculates the length of each street segment in feet, which is generally a block
long. As a check, the length of the first street segment is 663.647 feet. Computations will not
work for a few street segments that did not join with the MTFCC code table, so next you'll
select streets that did join, and calculations then will be performed for those streets only.

**8**   **Click Options > Select By Attributes, use the criterion** "StreetSpeeds.csv.SPEED" IS NOT
Null, **click Apply, and close the Select by Attributes window.** Some 18,015 out of 18,018
records are selected and joined correctly.

**9**   **Right-click FT_Minutes, click Field Calculator, enter the expression** [SFStreets.Length] /(88
× [StreetSpeeds.csv.SPEED]) **and click OK.** TF_Minutes for the first record is 0.301658.

**10** Use the same calculation for TF_Minutes.

**11** Click the Options button in the table and click Joins and Relates > Remove Join(s) > Remove All Joins.

**12** Click the Options button > Clear Selection, close the table, and save your map document.

## Create the network dataset

Here you'll run the New Network Dataset wizard.

**1** Open ArcCatalog. Click Customize > Extensions > Network Analyst > Close.

**2** Expand EsriPress > GIST1 > MyExercises > Chapter12, right-click SFStreets.shp, and click New Network Dataset. The New Network Dataset wizard opens. Note that the New Network Dataset option is available only for shapefile street layers and not feature datasets in file geodatabases.

**3** Click Next (accept the default name of SFStreets_ND) > Next (yes, you need to model turns and will use <Global Turns> as the turns source) > Next (use default connectivity) > Next (we don't have elevation data on overpasses so None is correct).

**4** Change Meters to Feet for the Length row in the New Network Dataset window.

**5** Click the button for Evaluators. Under Value, select Length, as shown in the image. Here you are selecting the Length field you computed in the previous exercise for use in measuring lengths of routes.

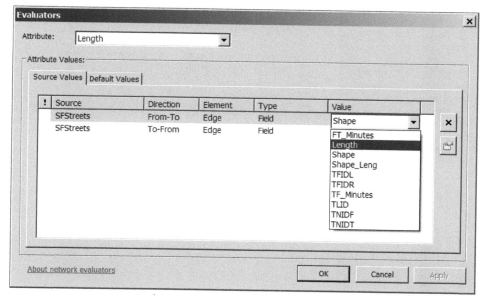

**6**  Do the same for the other row's Value (select Length) and click OK.

**7**  Double-click Minutes under Name and see that the wizard correctly identified FT_Minutes and TF_Minutes for Value fields. Click OK > Next > Next.

**8**  Click the Directions button and select FullName for the Name column on the Primary row of the lower panel.

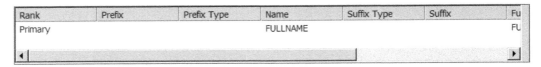

| Rank | Prefix | Prefix Type | Name | Suffix Type | Suffix | Fu |
|------|--------|-------------|----------|-------------|--------|-----|
| Primary | | | FULLNAME | | | FU |

**9**  Click OK > Next. The wizard lists all settings. If your settings don't match those shown in the image in bold font, use the back button and make corrections.

Turns:
  **<Global Turns>**
Connectivity:
  Group 1:
    Edge Connectivity:
     **SFStreets** (End Point)
Elevation Model: **None**
Attributes:
  Minutes:
    Usage Type: Cost
    Data Type: Double
    Units Type: **Minutes**
    Use by Default: True
    Source Attribute Evaluators:
     SFStreets (From-To): Field
      Language: VBScript
      Expression: **[FT_Minutes]**
     SFStreets (To-From): Field
      Language: VBScript
      Expression: **[TF_Minutes]**
    Default Attribute Evaluators:
     Default Edges: Constant = 0
     Default Junctions: Constant = 0
     Default Turns: Constant = 0
  Length:
    Usage Type: **Cost**

Data Type: Double
Units Type: **Feet**
Use by Default: False
Source Attribute Evaluators:
  SFStreets (From-To): Field
    Language: VBScript
    Expression: [Length]
  SFStreets (To-From): Field
    Language: VBScript
    Expression: [Length]
Default Attribute Evaluators:
  Default Edges: Constant = 0
  Default Junctions: Constant = 0
  Default Turns: Constant = 0
Directions:
  Directions Ready: Yes
  General Directions:
    Display Length Units: **Miles**
    Length Attribute: **Length**
    Time Attribute: **Minutes**
  Source Directions:
    SFStreets:
     Street Name Fields:
      Primary:
       Name: FULLNAME
       Full Name: **FULLNAME**

**10** Click Finish > Yes. That creates the network dataset SFStreets_ND.nd and street junctions as SFStreets_ND_Junctions.shp in your MyExercises/Chapter12 folder.

## Check global turn delays

As you will see, the Global Turn Delay Evaluator has well thought-out default parameters for traversing street intersections. Here you will only examine the default parameter values, but in the next Your Turn assignment, you'll change them.

**1** Double-click SFStreets_ND.nd in Catalog to open the Network Dataset Properties sheet.

**2** Click the Attributes tab and click the Evaluators button at the lower right.

**3** Select Minutes for Attribute.

**4** Click the Default Values tab, change Type for the Turn row from Constant to Global Turn Delay.

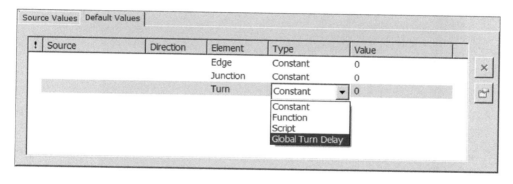

**5** Click the Evaluators Properties button and deselect the Only Show Turn Categories for Local Roads check box. The circle shows angles that are considered straight, reverse, or turns relative to driving north (or in any direction). The bottom panel has delay parameters in seconds for various ways to traverse an intersection, including turns. If you think about the times, they make sense; for example, the largest delay, 15 seconds, is from a local road onto a secondary road where you'd expect traffic on the secondary road to have the right of way and those on the local road to have a stop sign.

**6**   **Click Cancel > OK > OK.** Your network dataset is ready to use.

**7**   **Close Catalog.**

## YOUR TURN

**Add SFSstreets_ND.nd to your map in ArcMap, clicking No to add all feature classes that participate. Solve the traveling salesman problem for Run21097 using SFStreets_ND as the network dataset. Look back over the steps of tutorial 12-1, including (1) choose model and load stops, (2) make the kitchen, at 1117 Howard St, the first and last stop, and (3) set model parameters using Minutes as the impedance and reorder stops to find the optimal route while preserving the first and last stops. You should expect that total travel time will be less with this model run (and less accurate) because, for example, your street network dataset does not have restricted flows on one-way streets nor does it have restricted turns. In tutorial 12-1, with no barrier, the route was 19.8 miles long and took one hour. When you solve with your streets here, you'll find that the route is 18 miles long and takes 48 minutes.**

*(continued)* ➜

# YOUR TURN (continued)

Next, increase the global turn delay parameter values because they seem too small. That will increase travel time and perhaps change the route.

- Save your map document and close ArcMap. Open Catalog and double-click SFStreets_ND.
- Click the Attributes tab, double-click Minutes under Name.
- Click the Default Values tab.
- Click the Turn row of the Default Values tab on the Evaluators window to enable the Evaluators Property button.
- Click the Evaluators property button to open the Global Turn Delay Evaluator.
- Deselect the Only Show Turn Categories for Local Roads.
- Double all delay times (all times shown in the table).
- Click OK > OK > OK.

| Direction | Description | Seconds |
|---|---|---|
| Straight | From Local To Local Road Across No Roads | 0 |
| Straight | From Local To Local Road Across Local Road | 4 |
| Straight | From Local To Local Road Across Secondary or Primary Road | 30 |
| Straight | From Local To Secondary Road | 6 |
| Straight | From Secondary To Local Road | 6 |
| Straight | From Secondary To Secondary Road Across No Roads | 0 |
| Straight | From Secondary To Secondary Road Across Local Road | 1 |
| Straight | From Secondary To Secondary Road Across Secondary or Primary Road | 10 |
| Reverse | From Local To Local Road | 6 |
| Reverse | From Local To Secondary Road | 30 |
| Reverse | From Secondary To Local Road | 10 |
| Reverse | From Secondary To Secondary Road | 10 |
| Right Turn | From Local To Local Road | 4 |
| Right Turn | From Local To Secondary Road | 6 |
| Right Turn | From Secondary To Local Road | 4 |
| Right Turn | From Secondary To Secondary Road | 6 |
| Left Turn | From Local To Local Road | 4 |
| Left Turn | From Local To Secondary Road | 20 |
| Left Turn | From Secondary To Local Road | 10 |
| Left Turn | From Secondary To Secondary Road | 16 |

Open Tutorial12-2 in ArcMap and re-solve the model. You'll find that the route length stays about the same but travel time increases, closer to the time from tutorial 12-1. Of course there's no way to know if the increased turn times are valid for improving your network dataset or if the new time is closer to an hour by chance. When finished save your map document.

12

## TUTORIAL 12-3
# Creating travel polygons

During a city budget crisis, Pittsburgh city officials closed 16 of 32 public swimming pools to save costs. In tutorial 9-3, you created multiple ring buffers around open pools to estimate a gravity model curve of percentage of youths ages 5 to 17 who intended to use pools as a function of average distance from the nearest open pool. Intention to use a swimming pool is represented by owning a pool tag, which must be worn to be admitted to any Pittsburgh public pool. You found that the percentage of youths intending to use pools dropped quickly with distance in the expected manner. Here you will make a more accurate estimate of the same curve using multiple ring travel polygons based on street network travel to pools, assuming that youths take the shortest routes to pools.

## Get started

The map document you are about to open has the map layers of tutorial 9-3, which you used there to create multiple ring buffers around open swimming pools in Pittsburgh.

**1**   **Open Tutorial 12-3.mxd from the Maps folder.** The pool tag layer has a random sample of youths who owned pool tags. Each point is the residence of one pool tag owner. The block centroid layer has the population of youths, ages 5 to 17, by city block. New here is a street network dataset for Pittsburgh, PittsburghStreets_ND, built using the methods of tutorial 12-2 and with default global turn delays doubled, as you did in the previous Your Turn assignment.

**2**   Use the Pittsburgh bookmark.

**3**   Save the map document to the Chapter12 folder of MyExercises.

**4**   If necessary, click Customize > Extensions, click Network Analyst on, and click Close. Also click Customize > Toolbars > Network Analyst.

# Create multiple ring travel polygons

**1**   Click Selection > Select By Attributes, create the condition for pools, OPEN = 1, and click OK. That selects only the open pools so that when you load pools as locations for Network Analyst, only the open pools are included.

**2**   On the Network Analyst toolbar, click Network Analyst > New Service Area.

**3**   On the Network Analyst toolbar, click the Network Analyst window button.

**4**   Right-click Facilities and click Load Locations. Make sure that Pools is selected for the Load From field and that the Only load select rows box is checked and click OK. That loads the 16 open pools as facilities.

**5**   Click the Service Areas Properties button at the top right of the Network Analyst window and click the Analysis Settings tab.

**6**   Type or make selections as shown in the following figure with Length (Feet) for Impedance and Towards Facility marked for Direction (but don't click OK yet). The Default Breaks that you need to type are **2640**, **5280**, **7920**, **10560** in feet (corresponding to 0.5, 1.0, 1.5, and 2.0 miles). In tutorial 9-3, you used buffer radiuses of 0.25, 0.50, 0.75, 1.00, 1.50, and 2.00, but here 0.25 leads to too small sample sizes. Breaks above 2.0 miles also have sample sizes too small, and so you will not attempt them.

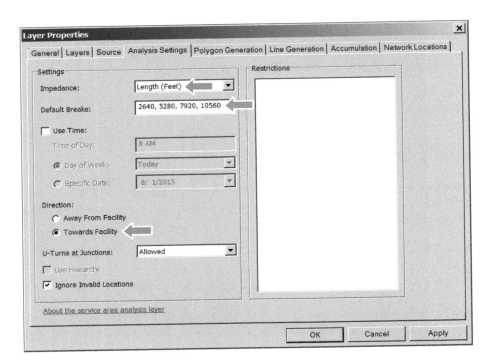

**7**  **Click the Polygon Generation tab of Layer Properties and make selections as shown, with Trim Polygons marked at 100 Feet, PittsburghStreets for Excluded Sources, Merge by break value marked, and Rings marked for Overlap Type.** These settings dissolve overlapping polygons of different pools as well as polygons for each interval of travel distances.

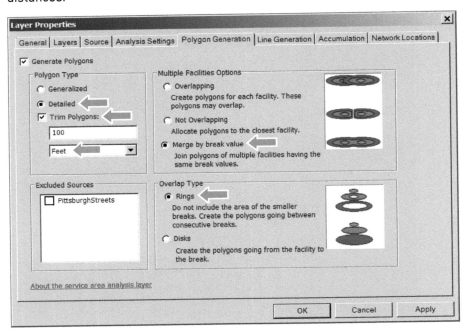

**8** **Click OK. Click the Solve button, symbolize the resulting polygons with the green-to-yellow-to-red color ramp, and turn off Pooltags for now.** Because Pittsburgh has three rivers and hilly topography, its streets are often irregular in shape—there's nothing remotely close to the grid system layout of streets in western states. As a result, there are many seemingly close locations to pools that are actually quite far as traveled through the street network. For example, examine the travel polygons of Jack Stack pool as shown in the two following images. The buffers of tutorial 9-3 extended all around the pool, but here you see that adjacent locations to the northwest of the pool are quite far when traveling by road.

## Spatially join travel polygons, pool tag owners, and youths

Next, you'll need to count youths with pool tags and sum up youth population, all by travel ring polygons. You'll follow the same workflow as in tutorial 9-3 for the same case, but for buffers: (1) spatially join the Pooltags layer to the travel polygons to produce polygons with the number of youths owning pool tags within each polygon; (2) spatially join the block centroids to the polygon layer created in step 1, which yields a new column with the population of youths in each polygon; (3) compute the percentage of youth population who have pool tags for each polygon; and (4) plot the percentages versus average distances from the nearest pool.

**1**   **Turn Pooltags back on.** It's helpful to see what you are joining.

**2**   **In the table of contents, right-click Polygons under Service Area, click Joins and Relates >**
       **Join, and type or make selections as shown in the following image, with Pooltags for no. 1,**
       **"Each polygon will be given a summary" marked for no. 2, and** Join1 **entered for no. 3.**

12

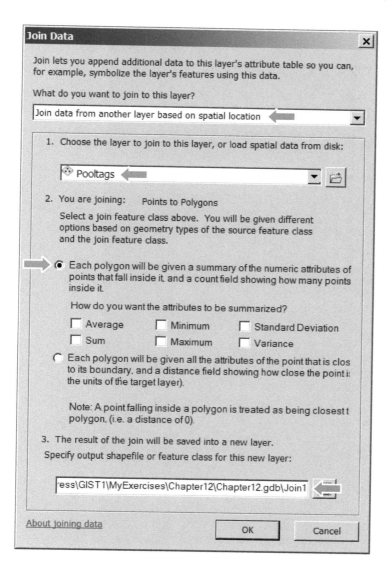

**3** Click OK. The new polygon layer has four records, one for each of the travel rings. The new attribute, Count_, has the number of pool tag owners from the random sample for each ring. Next, sum up the number of youths in each service area polygon.

**4** Right-click the new layer, Join1, click Joins and Relates > Join, and type or make selections as shown in the following image, with PittsburghBlockCentroids for no. 1, Sum checked for no. 2, and Join2 entered for no. 3.

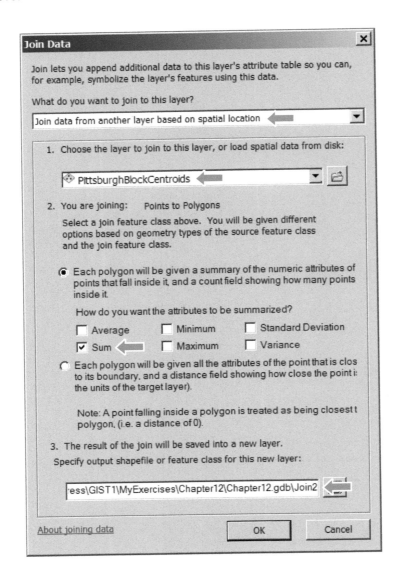

**5**   Click OK. Next you'll compute the percentage of youths in each ring. The pool tag data for youths is from a random sample. As explained in tutorial 9-3, these values need to be scaled up by a factor of 11.3 to estimate the total number of youths that had pool tags.

**6**   Open the attribute table of Join2 and create a new attribute named UseRate with the Float data type.

**7**   Open the Field Calculator for UseRate and compute the expression, $100 \times 11.3 \times$ [Count_]/ [Sum_AGE_5_17]. Count_ is the sample number of youths with pool tags and Sum_AGE_5_17 is the population of youths in each travel ring. You can see that nearly 100 percent of all youths are estimated to get pool tags within 0.5 miles of street travel of the nearest open pool, but then in the next half-mile travel, the percentage falls to 54 percent, and after that, it's flat at 44 percent out to two miles.

| Shape_Area | Name | Count_ | Sum_AGE_5_17 | UseRate |
|---|---|---|---|---|
| 273214287.120755 | 7920 - 10560 | 410 | 10504 | 44.10701 |
| 362025497.123333 | 5280 - 7920 | 541 | 13935 | 43.87011 |
| 316157103.574361 | 2640 - 5280 | 583 | 12213 | 53.9417 |
| 105910947.983058 | 0 - 2640 | 357 | 4071 | 99.09359 |

**8**   Save your map document.

## TUTORIAL 12-4
# Locating facilities

The Location-Allocation model of Network Analyst can choose the best among a set of potential facility sites to serve the clients within a region. Obviously, the set of potential facility sites needs to be equal to or larger than the number of facilities desired. For a particular facility type—say, for emergency care facilities or fast-food restaurants in a chain—there can be existing stores of the parent company or organization, plus stores of competitors. Demand for goods or services is from a point layer, which generally has centroids of blocks, census block groups, census tracts, or ZIP Codes with census population data. The census data can be total population, say, for emergency care facilities, or a particular demographic for fast-food restaurants (for example, population 5 to 49 years old). Several optimization models are available. You'll use the Maximize Attendance model, which assumes that the farther members of the target population reside from a facility, the lower the probability of them using it. You'll let the model choose the best 16 pools to open of the 32 public swimming pools in the Pittsburgh case study, and then compare the results to the pools actually kept open. The target demographic is youths ages 5 to 17 years old as represented by census data at block centroids. Of course, city officials used more criteria than just maximizing attendance, including the condition of pools and equity across areas of the city. So location-based access alone, as you're assessing here, is not necessarily the optimum sought by officials. Nevertheless, the solution that you'll find, if it had been available, should have been considered.

## Get started

The map document you are about to open has the map layers of tutorial 9-3, which you used to create multiple ring buffers around open swimming pools in Pittsburgh.

**1**   Open Tutorial 12-4.mxd from the Maps folder.

**2**    Save the map document to the Chapter12 folder of MyExercises.

**3**    If necessary, click Customize > Extensions, click Network Analyst on, and click Close. Also click Customize > Toolbars > Network Analyst.

## Load locations

You'll select the Location-Allocation model, and then load facility and demand points.

**1**    Click the Network Analyst on the Network Analyst toolbar and select New Location-Allocation.

**2**    Click the Network Analyst window button on the Network Analyst toolbar.

**3**    Right-click Facilities and click Load Locations. Make sure that Pools is selected for the Load From field and click OK. That loads the 32 pools as facilities.

**4**    Right-click Demand Points, click Load Locations, and make selections as shown in the image, with PittsburghBlockCentroids for Load From and Age_5_17 for Weight .

5   **Click OK and wait for the centroids to load.** All 5,041 block centroids of Pittsburgh load as demand points.

6   **Turn off Demand Points.** When you solve the facility-allocation problem, Network Analyst will draw lines between block centroids and the facilities to which they are assigned. So you still will be able to understand results.

## Set parameter values

1   **Click the Location-Allocation properties button in the Network Analyst window and click the Analysis Settings tab.**

12

**2** **Make selections as shown in the image, with Length (Feet) for Impedance and Demand to Facility for Travel From.** Next, you'll set parameters for a linear gravity model that declines from 100 percent of target population at zero distance from a pool to zero percent at 10,560 feet (2 miles). This function roughly approximates the gravity model estimated in tutorial 12-3.

**3** **Click the Advanced Settings tab and make or type selections as shown in the image, with Maximize Attendance for Problem Type,** 16 **for Facilities To Choose,** 10560 **for Impedance Cutoff, and Linear for Impedance Transformation. Click OK.**

## Solve the model

**1** Click the Solve button on the Network Analyst toolbar.

**2** In the table of contents, move the Pools layer to the top, turn off PittsburghStreets_ND, and give the Lines layer a width of 0.75. Now you can see which pools the city chose to keep open versus pools that Network Analyst chose to maximize attendance, assuming a linear gravity model. Network Analyst closed Sue Murray while opening Fowler in a one-for-one exchange. Those appear to be good choices by Network Analyst. South of the rivers, Sheraden, Westwood, Ormsby, Moore, and Phillips pools were kept open by the city, while Network Analyst closed Moore and Ormsby, and opened Beechwood, Warrington, Brookline, and Arlington. That's a net gain of two pools for this area, so the remaining center city region must have a net loss of two pools. The city had Ammon, Bloomfield, Highland, Schenley, Magee, Homewood, and East Hills open. Network Analyst closed Highland, East Hills, Schenley, and McBride pools while opening Paulson and Burgwin pools.

12

**3** Open the Facilities attribute table and get statistics for the DemandWeight attribute. Network Analyst estimates (with the sum) that 23,070 youths would get pool tags with this set of open pools and a linear gravity model.

**4** Close the Statistics of Facilities window and the attributes table. Save your map.

## TUTORIAL 12-5
# Routing vehicles from depots to demand points

Consumers want to buy locally grown produce (fruits, grains, and vegetables) in their grocery stores, but grocery stores cannot take deliveries from local farmers directly because of the small volume of produce a single farmer may erratically have versus the large volumes of produce reliably needed by the stores. It's too costly and there are too many uncertainties for large grocery stores to take delivery from many individual farmers. The solution is to provide aggregation depots, which take deliveries from farmers, large and small, and then deliver truckloads of produce to the grocery stores. That's just the sort of problem for the Vehicle Routing algorithm of Network Analyst; it minimizes costs for the depots of delivering the produce to grocery stores. As you will see, the Vehicle Routing algorithm is rich in terms of complexity handled and parameters you can set for minimizing routing costs.

## Get started

Suppose that the Food Hub Collaborative is looking for a produce aggregation depot in San Francisco and plans to use up to four dock-high delivery trucks, each with a payload capacity of 20,000 pounds, to supply San Francisco's over 40 major grocery stores with locally grown produce. For planning purposes, you'll assume that eventually all the grocery stores will be customers with typical demands based on sales volume. That way you can determine which of the two alternative depots is best and if four trucks is the right number.

**1**  Open Tutorial 12-5.mxd from the Maps folder and use the Service Area bookmark.

**2**  Save the map document to the Chapter12 folder of MyExercises.

**3**  If necessary, click Customize > Extensions, click Network Analyst on, and click Close. Also click Customize > Toolbars > Network Analyst.

## Add depots and orders

You'll select the Vehicle-Routing model, and then load facility and demand points. You'll select Sushi Umi and use it as the depot. Later, you'll try the US Trading Company as an alternative site.

**1**   Select the Sushi Umi depot with the Selection tool on the Tools toolbar.

**2**   On the Network Analyst toolbar, click Network Analyst and select New Vehicle Routing Problem. Open the Network Analyst window.

**3**   In the Network Analyst window, right-click Depots, click Load Locations, and type or make selections as shown in the image, with SFGroceryWarehouses **for Load From,** 9:00:00 **for TimeWindowStart1, and** 5:00:00 **for TimeWindowEnd1.** TimeWindowStart1 and TimeWindowEnd1 are the hours of operation for the depot. For Sushi Umi, it's 9:00 a.m. to 5:00 p.m.

**4**   Click OK.

**5** **Right-click Orders and click Load Locations. Select SFGroceryStores for the Load From field and complete the form by making selections as shown in the image, with** Address **for Name,** ServiceTime **for ServiceTIme,** TimeStart1 **for TimeWindowStart1,** TimeEnd1 **for TimeWindowEnd1, and** 0 **for MaxViolationTime1. Although the field is not shown in the image, scroll down and select Demand for DeliveryQuantities.** The solver needs unique names for each order point so you'll use Address, which is unique whereas Name repeats because chain stores have the same name. The ServiceTime field provides data on how long a stop is (in minutes) at a grocery store to get to the unloading dock, unload, and get back on the delivery route. TimeWindowStart1 and TimeWindowEnd1 provide the time window during which a grocery store accepts deliveries. Each store has a window of 9:00 a.m. to 5:00 p.m. The default value of 0 for MaxViolationTime1 indicates that time windows cannot be violated. Finally, DeliveryQuantities is the number of pounds of an order for each store.

**6** **Click OK.** That adds 43 orders to grocery stores to be on the route for deliveries.

# Add routes

Some details are needed for properties of routes and trucks; namely, the capacity and costs of operating trucks. Fortunately you can create one route (and truck), and then copy and paste it for the other three trucks because the Food Hub Collaborative would have four new, identical trucks. Then the Vehicle Routing Algorithm will create the four best routes to meet demand.

**1**  In the Network Analyst window, right-click Routes and click Add Item.

**2**  Select or type the properties from the following table (leaving properties omitted from the table with the default or Null value).

| Property | Value | Description |
|---|---|---|
| Name | Truck_1 | Name of the truck |
| StartDepotName | Sushi Umi | The truck starts at this depot |
| EndDepotName | Sushi Umi | The truck ends at this depot |
| EarliestStartTime | 8:00 AM | The truck can start loading at 8:00 AM |
| LatestStartTime | 8:00 AM | The truck must start at 8:00 AM |
| Capacities | 20000 | Capacity of the truck in pounds |
| CostPerUnitTime | 0.40 | The truck driver is paid $24 dollars per hour, so cost is $24/60 minutes = $0.40 per minute |
| CostPerUnitDistance | 1.25 | The cost of fuel, truck depreciation, maintenance, tires, etc. per mile |
| MaxTotalTime | 390 | The work shift for a driver is 8 hours, but total work time is 6.5 hours, or 390 minutes |

**3**  Click OK.

**4**  Right-click Truck_1 in the Network Analyst window and click Copy. Right-click Routes (1) and click Paste three times.

**5**  Edit the three copies of Truck _1 to rename them Truck_2, Truck _3, and Truck_4.

# Set properties for the routing problem

As is usual with Network Analyst, next you'll select analysis settings.

**1**  Click the Properties button in the Network Analyst window > Analysis Settings tab.

12

**2**  Type or make selections as shown in the image, with TravelTime (Minutes) for Time Attribute, Meters (Meters) for Distance Attribute, Monday for Day of Week, Not Allowed for U-Turns at Junctions, Straight Line for Output Shape Type, and Use Hierarchy unchecked.

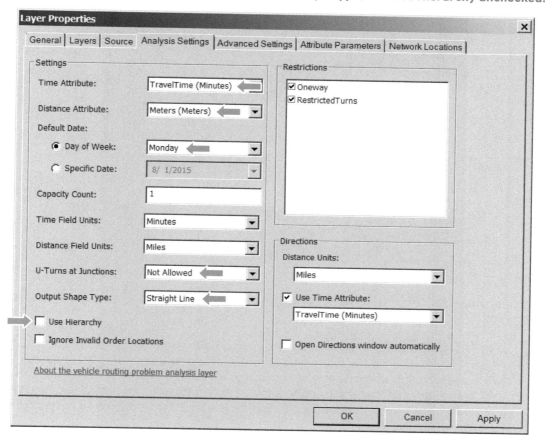

**3**  Click OK.

## Solve the routing problem

Now you'll see if four trucks can meet demand.

**1**  Click the Solve button on the Network Analyst toolbar.

**2** **Right-click Routes and symbolize with Categories > Unique values using Name for the Value field and width = 2 for line symbols.** Evidently, four trucks is enough to meet this level of demand, because all 43 grocery stores are allocated. Next, let's take a look at the level of usage for each truck and costs.

**3** **In the Network Analyst window under Routes, double-click Truck_1 to open its Properties window.** Scroll down to see the results of routing. This truck needed only 199.8 minutes of the 390 available in the workday, only drove about 6 miles for a total of 12.8 minutes of driving, and finished at 11:20 a.m. If you checked the other three trucks, you would see that they used 308, 350, and 371 minutes of the workday each. Maybe three trucks would be enough to do the job.

# Add route renewal

1   In the Network Analyst window, right-click and delete Truck_4.

2   Right-click Route Renewals (0) in the Network Analyst window and click Add item.

3   Make selections or type as follows. This allows Truck_1 to return to the depot to reload, in an estimated 30 minutes' time, and then go back out again to finish a route.

| Attribute | Value |
|-----------|-------|
| ObjectID | 1 |
| DepotName | Sushi Umi |
| RouteName | Truck_1 |
| ServiceTime | 30 |
| Sequences | <Null> |

4   Click OK. Likewise, add two more items to Route Renewals, including the other two trucks.

12

**5** **Click OK and re-solve.** That makes the three trucks nearly feasible with only two stores not getting deliveries. Truck driver 1 works 388 minutes, truck driver 2 works 355 minutes, and truck driver 3 works 382 minutes—getting close to the 390 minutes available in a regular workday. Not every store would have an order every day, so three trucks might be enough. The tight constraint is MaxTotalTime (390 minutes), so next try increasing that parameter to 7 hours per day x 60 minutes per hour = 420 minutes.

**6** **Under Routes in the Network Analyst window, double-click Truck_1 and set MaxTotalTime to 420. Likewise, do the same for the other two trucks and re-solve.** Now all grocery stores can have deliveries.

**7** **Save your map document.**

## ASSIGNMENT 12-1
# Geographic access to federally qualified health centers

Federally Qualified Health Centers (FQHCs) provide subsidized health services to underserved poor populations. There are 17 FQHCs in Pittsburgh, well-situated in poverty areas. The question addressed here is: How many people have easy geographic access to these centers? You'll compare two approaches to this problem: buffers versus service areas (travel-time polygons) computed using Network Analyst.

Suppose that 3 minutes' travel time or less to an FQHC is considered good geographic access. Given a street network, it's easy to estimate 3-minute travel-time polygons for the FQHCs.

For buffers, you need a buffer radius that roughly approximates three-minute maximum travel times. Suppose that non-rush-hour urban traffic—including stops, turns, and so forth—averages 15 to 20 miles per hour overall. Using 15 miles per hour (or 15/60 miles per minute), three minutes corresponds to a 0.75 mile radius buffer for straight-line travel from a residence on the buffer's perimeter to an FQHC. Of course, travel distance for residences on the buffer's perimeter where turns are necessary to get to a FQHC would be longer than 0.75 miles.

## Get set up

- Rename the folder \EsriPress\GIST1\MyAssignments\Chapter12\Assignment12-1YourName\ to your name. Store all files that you produce for this assignment in this folder.
- Create a new map document named **Assignment12-1YourName.mxd** with relative paths.

- Add the following map layers to your map and use cartographic principles for symbolization:
    - \EsriPress\GIST1\Data\NetworkAnalyst\PittsburghBlocks.shp—block polygons including POP10 = 2010 population attribute
    - \EsriPress\GIST1\Data\NetworkAnalyst\FQHC.shp—FQHC points with Name = name of FQHC attribute
    - \EsriPress\GIST1\Data\NetworkAnalyst\PittsburghStreets_ND—street network created from TIGER files using average travel times by street type
    - \EsriPress\GIST1\Data\Pittsburgh\City.gdb\Pittsburgh—outline of city
    - \EsriPress\GIST1\Data\Pittsburgh\City.gdb\Rivers—major rivers

Make *all* layers visible (including streets and layers you create) when zoomed to full extent. One exception is to turn off blocks.

## Requirements

Create 0.75-mile, merged buffers for the 17 FQHCs in Pittsburgh and create merged 3-minute travel-time polygons for the same FQHCs. Assume that any block that intersects a buffer or travel-time polygon has good access to one or more FQHCs (even though parts of some blocks are outside the buffers or travel-time polygons). Determine the population and percentage total population that has good geographic access to FQHCs based on buffers and travel-time polygons. Create a Word document named **Assignment12-1YourName.docx** that includes a layout of your finished map with FQHCs, buffers, travel-time polygons, streets, and other layers turned on, and includes the statistics requested. How good or bad are the buffer estimates compared with the travel-time estimates?

## ASSIGNMENT 12-2

# Analyze visits to the Phillips public pool in Pittsburgh

In this assignment, you analyze visits of youths to Phillips pool in Pittsburgh using service areas (travel-time polygons).

## Get set up

First, rename your assignment folder and create a map document.

- Rename the folder **\EsriPress\GIST1\MyAssignments\Chapter12\Assignment12-2YourName\** to your name. Store all files that you produce for this assignment in this folder.
- Create a new map document named **Assignment12-2YourName.mxd** with relative paths.

# Process the data

Add the following data to your map document:

- \EsriPress\GIST1\Data\NetworkAnalyst\PhillipsVisits.shp—data from a random sample of visitors to Phillips pool. Attributes include:
    - POOL = name of the pool that the pool tag owner stated that he/she intended to use
    - NOPERSONS = number of persons in the party or family for a visit
    - AGE = age of the person in the party included in the random sample
    - VISITS = number of visits that the person included in the random sample actually made to the pool during the summer studied
- Add the following definition query to PhillipsVisits so that you analyze only youths:
    - "AGE" <=17 AND "AGE" >= 5
- Symbolize with size-graduated point markers using the VISITS attribute.
    - \EsriPress\GIST1\Data\Pittsburgh\PublicPools.gdb\Pools—has open and closed Pittsburgh public pools. Symbolize with OPEN = 1 for open and OPEN = 0 for closed.
    - \EsriPress\GIST1\Data\NetworkAnalyst\PittsburghStreets_ND—street network created from TIGER files using average travel times by street type. Answer No to the question, "Do you want to add all features ...?"
    - EsriPress\GIST1\Data\Pittsburgh\City.gdb\Neighborhoods—for spatial context.

# Requirements

- Create multiple-ring service areas for the Phillips pool using impedances of minutes-travel-time to the pool. Create detailed polygon types and select the Rings overlap type. A major issue is to have a large-enough sample size for estimating behavior of each service area ring, so through trial and error, design a five-ring set of service area polygons so that each ring has at least 30 observations. Use whole numbers for impedances (such as 1, 2, 3, and so forth). The impedances do not have to follow a numeric sequence or be uniform. See the next step for getting statistics on resulting frequencies (counts) for each ring and average number of visits for a summer swimming season. *Hint:* Select Phillips pool in the Pools layer to restrict the service areas to that pool only.
- Spatially join PhillipsVisits to Polygons of the service area and include the average for all attributes, including avg_VISITS, for each buffer ring. Name the new layer **PhillipsServiceVisitsIntercept**.
- Create a scatter plot of average number of visits by youth versus average distance to buffers and include it in a layout with your map. You'll have to create a new field for average impedance; for example, if you have a ring corresponding to 3 to 5 minutes, use the average of 4 minutes to represent travel times in the ring. Add the graph to your layout by right-clicking it. (The graph can overlap the map somewhat.) What's the relationship of these two attributes? Include a text note on your layout describing the relationship. *Hint*: When you finish your graph, right-click it and click Copy as Graphic, so then you can paste it in your layout and it will maintain the graph's proportions (aspect ratio). If you right-click and click Add to Layout, ArcMap changes proportions.
- Export your map layout as **Assignment12-2YourName.jpg**.

## ASSIGNMENT 12-3

# Locate new farmers' markets in Washington, DC

Urban farmers' markets supply locally grown produce for communities and are especially important for poor neighborhoods that tend not to have grocery stores. This assignment uses data for Washington, DC, farmers' markets, plus block group data on populations with incomes below the poverty line. The objective is to select the best 5 of 15 potential sites for additional farmers' markets, keeping the existing markets and adding 5 more that maximize the number of low-income persons as potential customers.

## Get set up

- Rename the folder **\EsriPress\GIST1\MyAssignments\Chapter12\Assignment12-3YourName\** to your name. Store all files that you produce for this assignment in this folder.
- Create a new map document named **Assignment12-3YourName.mxd** with relative paths.

## Process the data

Add the following Washington, DC, data to your map document:

- \EsriPress\GIST1\Data\NetworkAnalyst\FarmersMarkets.shp—existing farmers' markets in Washington, DC.
- \EsriPress\GIST1\Data\NetworkAnalyst\XYNewLocs.shp—potential new locations for farmers' markets in Washington, DC.
- \EsriPress\GIST1\Data\NetworkAnalyst\DCPovBlkGrpCntrd.shp—2010 block group centroids with five-year 2013 ACS data on poverty.
    - Pop13 = population for which poverty status is known (577,244 total).
    - PopPov13 = population below poverty line (109,200 total).
- EsriPress\GIST1\Data\NetworkAnalyst\DCStreets_ND—Street network constructed from TIGER street centerlines following tutorial 12-2.
- \EsriPress\GIST1\Data\NetworkAnalyst\DC.shp—boundary of Washington, DC.

## Requirements

Using the Network Analyst toolbar, create a new Location-Allocation model. Add all 22 farmers' markets as required facility types and the 15 potential new locations as Candidate locations. *Hint*: In the Load Locations window, select the facility type in the rightmost column (Required or Candidate). Use minutes as impedance, and in the Advanced Settings of the Layer Properties window, use 27 for the facilities to choose (22 required and 5 of the 15 candidates).

We do not have a gravity model calibrated with customer data and so must exercise judgment for the falloff of farmer market use with travel time, along with a sensitivity analysis. Use Linear for the Impedance Transformation with three alternative impedance cutoffs for three runs of the model: 15, 20, and 25 minutes, to see if the solution of chosen potential sites is robust (does not vary with variation in cutoffs). Recall that for a cutoff, the model allocates 100 percent of population as users at 0 minutes, and then linearly decreases users to 0 percent at the cutoff. Here we're guessing a linear decline with 15 minutes for cutoff as conservative, 20 minutes as expected, and 25 minutes as optimistic. Run the model with each of these times. Create and fill out a table in Word similar to the following, in which you enter the names of five chosen sites for each cutoff and highlight any differences. Also include the sum of DemandWeight for each cutoff (and both required and chosen sites), which is the estimate of total users.

| | 15 Minute Cutoff | 20 Minute Cutoff | 25 Minute Cutoff |
|---|---|---|---|
| Chosen | | | |
| Chosen | | | |
| Chosen | | | |
| Chosen | | | |
| Chosen | | | |
| Estimated Users | | | |

Before you turn in your assignment, run the model for the 20-minute cutoff so that its solution is in your map document. Create a Word document with a title, your name, a table as shown, and a layout of your map showing the solution, including allocations of block group centroids to farmers' market sites.

12

# Data source credits

## CMUCampus

\EsriPress\GIST1\Data\CMUCampus\CampusMap.dwg, courtesy of Carnegie Mellon University.

\EsriPress\GIST1\Data\CMUCampus\CampusMap.dwg\Annotation, courtesy of Carnegie Mellon University.

\EsriPress\GIST1\Data\CMUCampus\CampusMap.dwg\MultiPatch, courtesy of Carnegie Mellon University.

\EsriPress\GIST1\Data\CMUCampus\CampusMap.dwg\Point, courtesy of Carnegie Mellon University.

\EsriPress\GIST1\Data\CMUCampus\CampusMap.dwg\Polygon, courtesy of Carnegie Mellon University.

\EsriPress\GIST1\Data\CMUCampus\CampusMap.dwg\Polyline, courtesy of Carnegie Mellon University.

\EsriPress\GIST1\Data\CMUCampus\Walkways.dwg (from CampusMap.dwg), courtesy of Carnegie Mellon University.

\EsriPress\GIST1\Data\CMUCampus\Walkways.dwg\Annotation, courtesy of Carnegie Mellon University.

\EsriPress\GIST1\Data\CMUCampus\Walkways.dwg\MultiPatch, courtesy of Carnegie Mellon University.

\EsriPress\GIST1\Data\CMUCampus\Walkways.dwg\Point, courtesy of Carnegie Mellon University.

\EsriPress\GIST1\Data\CMUCampus\Walkways.dwg\Polygon, courtesy of Carnegie Mellon University.

\EsriPress\GIST1\Data\CMUCampus\Walkways.dwg\Polyline, courtesy of Carnegie Mellon University.

\EsriPress\GIST1\Data\CMUCampus\CMUCampus.tif, image courtesy of US Geological Survey, Department of the Interior/USGS.

\EsriPress\GIST1\Data\CMUCampus\CMUCampus.tif\Band_1, image courtesy of US Geological Survey, Department of the Interior/USGS.

\EsriPress\GIST1\Data\CMUCampus\CMUCampus.tif\Band_2, image courtesy of US Geological Survey, Department of the Interior/USGS.

\EsriPress\GIST1\Data\CMUCampus\CMUCampus.tif\Band_3, image courtesy of US Geological Survey, Department of the Interior/USGS.

\EsriPress\GIST1\Data\CMUCampus\HBH.shp, courtesy of Carnegie Mellon University.

\EsriPress\GIST1\Data\CMUCampus\MainCampusBuildings.shp (from CampusMap.dwg), courtesy of Carnegie Mellon University.

## DataFiles

\EsriPress\GIST1\Data\DataFiles\AllCoEdAttain.xlsx, courtesy of US Census Bureau.

\EsriPress\GIST1\Data\DataFiles\AllCoEdAttain.xlsc\HighEdAttainment$, courtesy of US Census Bureau.

\EsriPress\GIST1\Data\DataFiles\AlleghenyCountyBlockGroups.shp, courtesy of US Census Bureau.

\EsriPress\GIST1\Data\DataFiles\AlleghenyCountyTracts.shp, courtesy of US Census Bureau.

\EsriPress\GIST1\Data\DataFiles\AsianTop10States.xlsx, courtesy of US Census Bureau and Wilpen Gorr.

\EsriPress\GIST1\Data\DataFiles\AsianTop10States.xlsx\AsianTop10States$, courtesy of US Census Bureau and Wilpen Gorr.

\EsriPress\GIST1\Data\DataFiles\AutoTheftCrimeSeries.shp, courtesy of Department of City Planning, City of Pittsburgh.

\EsriPress\GIST1\Data\DataFiles\DCTract.shp, courtesy of US Census Bureau.

\EsriPress\GIST1\Data\DataFiles\Earthquakes.dbf, from Esri Data & Maps 2004, courtesy of National Atlas of the United States, USGS.

\EsriPress\GIST1\Data\DataFiles\FishnetCalculations.xlsx, by Wilpen Gorr.

\EsriPress\GIST1\Data\DataFiles\PghCrosswalks.dbf, courtesy of Department of City Planning, City of Pittsburgh, and Wilpen Gorr.

\EsriPress\GIST1\Data\DataFiles\PghTracts.shp, courtesy of US Census Bureau.

\EsriPress\GIST1\Data\DataFiles\PittsburghSchools.csv, from Esri Data & Maps 2010, courtesy of US Census Bureau.

\EsriPress\GIST1\Data\DataFiles\Statistics.txt, courtesy of US Census Bureau and Wilpen Gorr.

\EsriPress\GIST1\Data\DataFiles\tl_2010_42003_tract10.shp, courtesy of US Census Bureau.

\EsriPress\GIST1\Data\DataFiles\Tornadoes2000.shp, courtesy of National Atlas of the United States, downloaded from NationalAtlas.gov (http://www.nationalatlas.gov/atlasftp.html#tornadx).

\EsriPress\GIST1\Data\DataFiles\Neighborhoods.zip, courtesy of Department of City Planning, City of Pittsburgh.

## MaricopaCounty

\EsriPress\GIST1\Data\MaricopaCounty\CensusData.xlsx, courtesy of US Census Bureau.

\EsriPress\GIST1\Data\MaricopaCounty\CensusData.xlsx\CensusData,courtesy of US Census Bureau.

\EsriPress\GIST1\Data\MaricopaCounty\CensusData.xlsx\CensusData$, courtesy of US Census Bureau.

\EsriPress\GIST1\Data\MaricopaCounty\tl_2010_04013_cousub10.shp, courtesy of US Census Bureau.

\EsriPress\GIST1\Data\MaricopaCounty\tl_2010_04013_tract10.shp, courtesy of US Census Bureau.

## EastLiberty

\EsriPress\GIST1\Data\Pittsburgh\EastLiberty\Building\Annotation, courtesy of Department of City Planning, City of Pittsburgh.

\EsriPress\GIST1\Data\Pittsburgh\EastLiberty\Building\Arc, courtesy of Department of City Planning, City of Pittsburgh.

\EsriPress\GIST1\Data\Pittsburgh\EastLiberty\Building\Label, courtesy of Department of City Planning, City of Pittsburgh.

\EsriPress\GIST1\Data\Pittsburgh\EastLiberty\Building\Polygon, courtesy of Department of City Planning, City of Pittsburgh.

\EsriPress\GIST1\Data\Pittsburgh\EastLiberty\Building\Tic, courtesy of Department of City Planning, City of Pittsburgh.

\EsriPress\GIST1\Data\Pittsburgh\EastLiberty\Curbs\Annotation, courtesy of Department of City Planning, City of Pittsburgh.

\EsriPress\GIST1\Data\Pittsburgh\EastLiberty\Curbs\Arc, courtesy of Department of City Planning, City of Pittsburgh.

\EsriPress\GIST1\Data\Pittsburgh\EastLiberty\Curbs\Tic, courtesy of Department of City Planning, City of Pittsburgh.

\EsriPress\GIST1\Data\Pittsburgh\EastLiberty\EastLib\Annotation, courtesy of Department of City Planning, City of Pittsburgh.

\EsriPress\GIST1\Data\Pittsburgh\EastLiberty\EastLib\Arc, courtesy of Department of City Planning, City of Pittsburgh.

\EsriPress\GIST1\Data\Pittsburgh\EastLiberty\EastLib\Label, courtesy of Department of City Planning, City of Pittsburgh.

\EsriPress\GIST1\Data\Pittsburgh\EastLiberty\EastLib\Polygon, courtesy of Department of City Planning, City of Pittsburgh.

\EsriPress\GIST1\Data\Pittsburgh\EastLiberty\EastLib\Tic, courtesy of Department of City Planning, City of Pittsburgh.

\EsriPress\GIST1\Data\Pittsburgh\EastLiberty\Parcel\Annotation, courtesy of Department of City Planning, City of Pittsburgh.

\EsriPress\GIST1\Data\Pittsburgh\EastLiberty\Parcel\Arc, courtesy of Department of City Planning, City of Pittsburgh.

\EsriPress\GIST1\Data\Pittsburgh\EastLiberty\Parcel\Label, courtesy of Department of City Planning, City of Pittsburgh.

\EsriPress\GIST1\Data\Pittsburgh\EastLiberty\Parcel\Polygon, courtesy of Department of City Planning, City of Pittsburgh.

\EsriPress\GIST1\Data\Pittsburgh\EastLiberty\Parcel\Tic, courtesy of Department of City Planning, City of Pittsburgh.

## Shapefiles

\EsriPress\GIST1\Data\Pittsburgh\Shapefiles\Larcenies.shp, courtesy of Pittsburgh Bureau of Police.

\EsriPress\GIST1\Data\Pittsburgh\Shapefiles\Parks.shp, courtesy of Southwestern Pennsylvania Commission.

\EsriPress\GIST1\Data\Pittsburgh\Shapefiles\PittsburghSeriousCrimes2008.shp, courtesy of Pittsburgh Bureau of Police.

\EsriPress\GIST1\Data\Pittsburgh\Shapefiles\PovertyTracts.xlsx, courtesy of US Census Bureau.

\EsriPress\GIST1\Data\Pittsburgh\Shapefiles\PovertyTracts.xlsx\PghPovertyTracts$, courtesy of US Census Bureau.

## CBD

\EsriPress\GIST1\Data\Pittsburgh\CBD.gdb\BldgNameAlias, courtesy of Wilpen Gorr, Carnegie Mellon University.

\EsriPress\GIST1\Data\Pittsburgh\CBD.gdb\Bldgs, courtesy of Department of City Planning, City of Pittsburgh.

\EsriPress\GIST1\Data\Pittsburgh\CBD.gdb\Clients, courtesy of Kristen Kurland, Carnegie Mellon University.

\EsriPress\GIST1\Data\Pittsburgh\CBD.gdb\Curbs, courtesy of Department of City Planning, City of Pittsburgh.

\EsriPress\GIST1\Data\Pittsburgh\CBD.gdb\Histpnts, courtesy of Department of City Planning, City of Pittsburgh.

\EsriPress\GIST1\Data\Pittsburgh\CBD.gdb\Histsites, courtesy of Department of City Planning, City of Pittsburgh.

\EsriPress\GIST1\Data\Pittsburgh\CBD.gdb\Outline, courtesy of Department of City Planning, City of Pittsburgh.

\EsriPress\GIST1\Data\Pittsburgh\CBD.gdb\Parcels, courtesy of Department of City Planning, City of Pittsburgh.

\EsriPress\GIST1\Data\Pittsburgh\CBD.gdb\Streets, courtesy of Department of City Planning, City of Pittsburgh.

## City

\EsriPress\GIST1\Data\Pittsburgh\City.gdb\ATMRobberies, courtesy of Pittsburgh Bureau of Police.

\EsriPress\GIST1\Data\Pittsburgh\City.gdb\ImmigrantBusinesses, courtesy of Kristen Kurland, Carnegie Mellon University.

\EsriPress\GIST1\Data\Pittsburgh\City.gdb\Neighborhoods, courtesy of Department of City Planning, City of Pittsburgh.

\EsriPress\GIST1\Data\Pittsburgh\City.gdb\PghStreets, courtesy of Department of City Planning, City of Pittsburgh.

\EsriPress\GIST1\Data\Pittsburgh\City.gdb\PghTracts, courtesy of US Census Bureau.

\EsriPress\GIST1\Data\Pittsburgh\City.gdb\Pittsburgh, courtesy of Southwestern Pennsylvania Commission.

\EsriPress\GIST1\Data\Pittsburgh\City.gdb\Rivers, courtesy of US Census Bureau.

\EsriPress\GIST1\Data\Pittsburgh\City.gdb\Schools, courtesy of Department of City Planning, City of Pittsburgh.

\EsriPress\GIST1\Data\Pittsburgh\City.gdb\Zoning, courtesy of Department of City Planning, City of Pittsburgh.

## Midhill

\EsriPress\GIST1\Data\Pittsburgh\Midhill.gdb\Bldgs, courtesy of Department of City Planning, City of Pittsburgh.

\EsriPress\GIST1\Data\Pittsburgh\Midhill.gdb\CADCalls, courtesy of Department of City Planning, City of Pittsburgh.

\EsriPress\GIST1\Data\Pittsburgh\Midhill.gdb\CommercialProperties, courtesy of Department of City Planning, City of Pittsburgh.

\EsriPress\GIST1\Data\Pittsburgh\Midhill.gdb\Curbs, courtesy of Department of City Planning, City of Pittsburgh.

\EsriPress\GIST1\Data\Pittsburgh\Midhill.gdb\Outline, courtesy of Department of City Planning, City of Pittsburgh.

\EsriPress\GIST1\Data\Pittsburgh\Midhill.gdb\Schools, courtesy of National Agriculture Imagery Program (NAIP), US Department of Agriculture, Farm Service Agency.

\EsriPress\GIST1\Data\Pittsburgh\Midhill.gdb\Streets, courtesy of Department of City Planning, City of Pittsburgh.

## Phipps

\EsriPress\GIST1\Data\Pittsburgh\Phipps.gdb\Bldgs, courtesy of Department of City Planning, City of Pittsburgh.

\EsriPress\GIST1\Data\Pittsburgh\Phipps.gdb\Curbs, courtesy of Department of City Planning, City of Pittsburgh.

\EsriPress\GIST1\Data\Pittsburgh\Phipps.gdb\Image\Band_1, courtesy of US Geological Survey, Department of the Interior/USGS.

\EsriPress\GIST1\Data\Pittsburgh\Phipps.gdb\Image\Band_2, courtesy of US Geological Survey, Department of the Interior/USGS.

\EsriPress\GIST1\Data\Pittsburgh\Phipps.gdb\Image\Band_3, courtesy of US Geological Survey, Department of the Interior/USGS.

\EsriPress\GIST1\Data\Pittsburgh\Phipps.gdb\PhippsAddition, digitized by Kristen Kurland, Carnegie Mellon University.

\EsriPress\GIST1\Data\Pittsburgh\Phipps.gdb\Topo, courtesy of Department of City Planning, City of Pittsburgh.

\EsriPress\GIST1\Data\Pittsburgh\PublicPools.gdb\JackStackVisits, courtesy of Pittsburgh Citiparks Department.

\EsriPress\GIST1\Data\Pittsburgh\PublicPools.gdb\PittsburghBlockCentroids, courtesy of US Census Bureau and Wilpen Gorr.

\EsriPress\GIST1\Data\Pittsburgh\PublicPools.gdb\Pools, courtesy of Pittsburgh Citiparks Department.

\EsriPress\GIST1\Data\Pittsburgh\PublicPools.gdb\Pooltags, courtesy of Pittsburgh Citiparks Department.

## Zone2

\EsriPress\GIST1\Data\Pittsburgh\Zone2.gdb\Outline, courtesy of Department of City Planning, City of Pittsburgh.

\EsriPress\GIST1\Data\Pittsburgh\Zone2.gdb\Streets, courtesy of US Census Bureau.

## RochesterNY

\EsriPress\GIST1\Data\RochesterNY\LakePrecinct.gdb\lakebars, courtesy of Rochester, New York, Police Department.

\EsriPress\GIST1\Data\RochesterNY\LakePrecinct.gdb\LakeBusinesses, courtesy of InfoUSA.

\EsriPress\GIST1\Data\RochesterNY\LakePrecinct.gdb\lakecarbeats, courtesy of Rochester, New York, Police Department.

\EsriPress\GIST1\Data\RochesterNY\LakePrecinct.gdb\lakeprecinct, courtesy of Rochester, New York, Police Department.

\EsriPress\GIST1\Data\RochesterNY\LakePrecinct.gdb\lakestreets, courtesy of Rochester, New York, Police Department.

\EsriPress\GIST1\Data\RochesterNY\RochesterPolice.gdb\business, courtesy of Rochester, New York, Police Department.

\EsriPress\GIST1\Data\RochesterNY\RochesterPolice.gdb\carbeats, courtesy of Rochester, New York, Police Department.

\EsriPress\GIST1\Data\RochesterNY\RochesterPolice.gdb\SIC, courtesy of US Census Bureau.

\EsriPress\GIST1\Data\RochesterNY\RochesterPolice.gdb\streets, courtesy of US Census Bureau.

## SpatialAnalyst

\EsriPress\GIST1\Data\SpatialAnalyst\LandUse\LandUse_Pgh.tif, image courtesy of US Geological Survey, Department of the Interior/USGS.

\EsriPress\GIST1\Data\SpatialAnalyst\LandUse\LandUse_Pgh.prj, courtesy of Wilpen Gorr.

\EsriPress\GIST1\Data\SpatialAnalyst\SpatialAnalyst.gdb\AllCoBlkGrps, courtesy of US Census Bureau TIGER.

\EsriPress\GIST1\Data\SpatialAnalyst\SpatialAnalyst.gdb\AllCoBlocks, courtesy of US Census Bureau TIGER.

\EsriPress\GIST1\Data\SpatialAnalyst\SpatialAnalyst.gdb\NED, courtesy of US Geological Survey, Department of the Interior/USGS.

\EsriPress\GIST1\Data\SpatialAnalyst\SpatialAnalyst.gdb\OHCA, courtesy of Children's Hospital of Pittsburgh.

\EsriPress\GIST1\Data\SpatialAnalyst\SpatialAnalyst.gdb\PennHills, courtesy of US Census Bureau.

\EsriPress\GIST1\Data\SpatialAnalyst\SpatialAnalyst.gdb\Pittsburgh, courtesy of US Census Bureau.

\EsriPress\GIST1\Data\SpatialAnalyst\SpatialAnalyst.gdb\Rivers, courtesy of US Census Bureau.

\EsriPress\GIST1\Data\SpatialAnalyst\SpatialAnalyst.gdb\Shaler, courtesy of US Census Bureau.

\EsriPress\GIST1\Data\SpatialAnalyst\SpatialAnalyst.gdb\ZoningCommercialBuffer, courtesy of Department of City Planning, City of Pittsburgh.

\EsriPress\GIST1\Data\SpatialAnalyst\LandUse.lyr, layer file courtesy of Wilpen Gorr with symbolization for LandUse_Pgh.tif.

## 3DAnalyst

\EsriPress\GIST1\Data\3DAnalyst.gdb\Bldgs, courtesy of Department of City Planning, City of Pittsburgh.

\EsriPress\GIST1\Data\3DAnalyst.gdb\Curbs, courtesy of Department of City Planning, City of Pittsburgh.

\EsriPress\GIST1\Data\3DAnalyst.gdb\Image\Band_1, courtesy of US Geological Survey, Department of the Interior/USGS.

\EsriPress\GIST1\Data\3DAnalyst.gdb\Image\Band_2, courtesy of US Geological Survey, Department of the Interior/USGS.

\EsriPress\GIST1\Data\3DAnalyst.gdb\NorthShoreTrees, courtesy of Kristen Kurland, Carnegie Mellon University.

\EsriPress\GIST1\Data\3DAnalyst.gdb\NorthShoreVehicles, courtesy of Kristen Kurland, Carnegie Mellon University.

\EsriPress\GIST1\Data\3DAnalyst.gdb\ProposedBldgs, courtesy of Kristen Kurland, Carnegie Mellon University.

\EsriPress\GIST1\Data\3DAnalyst.gdb\Rivers, courtesy of US Census Bureau.

\EsriPress\GIST1\Data\3DAnalyst.gdb\StudyArea, courtesy of Kristen Kurland, Carnegie Mellon University.

\EsriPress\GIST1\Data\3DAnalyst.gdb\Topo, courtesy of Department of City Planning, City of Pittsburgh.

## AlleghenyCounty

\EsriPress\GIST1\Data\AlleghenyCounty.gdb\CountySchools, from Esri Data & Maps 2010, courtesy of US Census Bureau.

\EsriPress\GIST1\Data\AlleghenyCounty.gdb\Munic, courtesy of Southwestern Pennsylvania Commission.

\EsriPress\GIST1\Data\AlleghenyCounty.gdb\Parks, courtesy of Southwestern Pennsylvania Commission.

\EsriPress\GIST1\Data\AlleghenyCounty.gdb\Rivers, courtesy of US Census Bureau.

\EsriPress\GIST1\Data\AlleghenyCounty.gdb\Tracts, courtesy of US Census Bureau.

\EsriPress\GIST1\Data\AlleghenyCounty.gdb\Tracts2010, courtesy of US Census Bureau.

## NYC

\EsriPress\GIST1\Data\NYC.gdb\Boroughs. Material titled GIS Data used with permission of New York City Department of City Planning. All rights reserved.

\EsriPress\GIST1\Data\NYC.gdb\BronxWater, courtesy of US Census Bureau.

\EsriPress\GIST1\Data\NYC.gdb\BronxWaterfrontParks. Material titled GIS Data used with permission of New York City Department of City Planning. All rights reserved.

\EsriPress\GIST1\Data\NYC.gdb\BrooklynWaterfrontParks. Material titled GIS Data used with permission of New York City Department of City Planning. All rights reserved.

\EsriPress\GIST1\Data\NYC.gdb\CouncilDistricts. Material titled GIS Data used with permission of New York City Department of City Planning. All rights reserved.

\EsriPress\GIST1\Data\NYC.gdb\Facilities. Material titled GIS Data used with permission of New York City Department of City Planning. All rights reserved.

\EsriPress\GIST1\Data\NYC.gdb\FireCompanies. Material titled GIS Data used with permission of New York City Department of City Planning. All rights reserved.

\EsriPress\GIST1\Data\NYC.gdb\HealthAreas. Material titled GIS Data used with permission of New York City Department of City Planning. All rights reserved.

\EsriPress\GIST1\Data\NYC.gdb\HealthCenters. Material titled GIS Data used with permission of New York City Department of City Planning. All rights reserved.

\EsriPress\GIST1\Data\NYC.gdb\KingsWater, courtesy of US Census Bureau.

\EsriPress\GIST1\Data\NYC.gdb\ManhattanStreets. Material titled GIS Data used with permission of New York City Department of City Planning. All rights reserved.

\EsriPress\GIST1\Data\NYC.gdb\ManhattanTracts, courtesy of US Census Bureau.

\EsriPress\GIST1\Data\NYC.gdb\ManhattanWaterfrontParks. Material titled GIS Data used with permission of New York City Department of City Planning. All rights reserved.

\EsriPress\GIST1\Data\NYC.gdb\MetroRoads, from Esri Data & Maps 2004, courtesy of US Census Bureau.

\EsriPress\GIST1\Data\NYC.gdb\Neighborhoods. Material titled GIS Data used with permission of New York City Department of City Planning. All rights reserved.

\EsriPress\GIST1\Data\NYC.gdb\NewYorkWater, courtesy of US Census Bureau.

\EsriPress\GIST1\Data\NYC.gdb\PolicePrecincts. Material titled GIS Data used with permission of New York City Department of City Planning. All rights reserved.

\EsriPress\GIST1\Data\NYC.gdb\QueensWater, courtesy of US Census Bureau.

\EsriPress\GIST1\Data\NYC.gdb\QueensWaterfrontParks. Material titled GIS Data used with permission of New York City Department of City Planning. All rights reserved.

\EsriPress\GIST1\Data\NYC.gdb\RichmondWater, courtesy of US Census Bureau.

\EsriPress\GIST1\Data\NYC.gdb\SchoolDistricts. Material titled GIS Data used with permission of New York City Department of City Planning. All rights reserved.

\EsriPress\GIST1\Data\NYC.gdb\StatenIslandWaterfrontParks. Material titled GIS Data used with permission of New York City Department of City Planning. All rights reserved.

\EsriPress\GIST1\Data\NYC.gdb\Water, courtesy of US Census Bureau.

\EsriPress\GIST1\Data\NYC.gdb\ZoningLandUse. Material titled GIS Data used with permission of New York City Department of City Planning. All rights reserved.

## UnitedStates

\EsriPress\GIST1\Data\UnitedStates.gdb\CACounties, from Esri Data & Maps 2010, courtesy of ArcUSA, US Census Bureau, Esri (POP2010 field).

\EsriPress\GIST1\Data\UnitedStates.gdb\CAOrangeCountyTracts, from Esri Data & Maps 2010, courtesy of Tele Atlas, US Census Bureau, Esri (POP2010 field).

\EsriPress\GIST1\Data\UnitedStates.gdb\COCounties, Esri Data & Maps 2010, courtesy of ArcUSA, US Census Bureau, Esri (POP2010 field).

\EsriPress\GIST1\Data\UnitedStates.gdb\Congress111, courtesy of US Census Bureau.

\EsriPress\GIST1\Data\UnitedStates.gdb\COStreets, Esri Data & Maps 2010, courtesy of ArcUSA, US Census Bureau, Esri (POP2010 field).

\EsriPress\GIST1\Data\UnitedStates.gdb\COStreets1, from Esri Data & Maps 2010, courtesy of ArcUSA, US Census Bureau, Esri (POP2010 field).

\EsriPress\GIST1\Data\UnitedStates.gdb\COStreets2, from Esri Data & Maps 2010, courtesy of ArcUSA, US Census Bureau, Esri (POP2010 field).

\EsriPress\GIST1\Data\UnitedStates.gdb\COUrban1, courtesy of US Census Bureau.

\EsriPress\GIST1\Data\UnitedStates.gdb\COUrban2, courtesy of US Census Bureau.

\EsriPress\GIST1\Data\UnitedStates.gdb\Water1, courtesy of US Census Bureau.

\EsriPress\GIST1\Data\United States.gdb\Water2, courtesy of US Census Bureau.

\EsriPress\GIST1\Data\UnitedStates.gdb\HHWZIPCodes, courtesy of Pennsylvania Resources Council.

\EsriPress\GIST1\Data\UnitedStates.gdb\MilitaryBnd, courtesy of DISDI, downloaded from Data.gov (http://www.data.gov/geodata/g735939/), DISDI. Data.gov and the US federal government cannot vouch for the data or analyses derived from this data after the data has been retrieved from Data.gov.

\EsriPress\GIST1\Data\UnitedStates.gdb\MilitaryPt, courtesy of DISDI, downloaded from Data.gov (http://www.data.gov/geodata/g736859/), DISDI. Data.gov and the US federal government cannot vouch for the data or analyses derived from this data after the data has been retrieved from Data.gov.

\EsriPress\GIST1\Data\UnitedStates.gdb\PACities, from Esri Data & Maps 2010, courtesy of National Atlas of the United States.

\EsriPress\GIST1\Data\UnitedStates.gdb\PACounties, from Esri Data & Maps 2010, courtesy of ArcUSA, US Census Bureau, Esri (POP2010 field).

\EsriPress\GIST1\Data\UnitedStates.gdb\PAPublicSchools, from Esri Data & Maps 2010, courtesy of Tele Atlas (TomTom), US Census Bureau, Esri (POP2010 field).

\EsriPress\GIST1\Data\UnitedStates.gdb\PATracts, from Esri Data & Maps 2010,courtesy of Tele Atlas (TomTom), US Census Bureau, Esri (POP2010 field).

\EsriPress\GIST1\Data\UnitedStates.gdb\PAZIP, courtesy of Tele Atlas (TomTom), Esri (POP2010 field).

\EsriPress\GIST1\Data\UnitedStates.gdb\USCities, from Esri Data & Maps 2010, courtesy of US Census Bureau.

\EsriPress\GIST1\Data\UnitedStates.gdb\USCities_dtl, from Esri Data & Maps 2010, courtesy of US Census Bureau.

\EsriPress\GIST1\Data\UnitedStates.gdb\USCounties, from Esri Data & Maps 2010, courtesy of ArcUSA, US Census Bureau, Esri (POP2010 field).

\EsriPress\GIST1\Data\UnitedStates.gdb\USStates, from Esri Data & Maps 2010, courtesy of ArcUSA, US Census Bureau, Esri (POP2010 field).

## World

\EsriPress\GIST1\Data\World.gdb\Country, courtesy of *ArcWorld Supplement*.

\EsriPress\GIST1\Data\World.gdb\Ocean, from Esri Data & Maps 2004, courtesy of Esri.

## Network Analyst

\EsriPress\GIST1\Data\NetworkAnalyst\DC.shp, courtesy of US Census Bureau.

\EsriPress\GIST1\Data\NetworkAnalyst\DCPovBlkGrpCntrd.shp, courtesy of US Census Bureau and Wilpen Gorr.

\EsriPress\GIST1\Data\NetworkAnalyst\DCStreets.shp, courtesy of US Census Bureau and Wilpen Gorr.

\EsriPress\GIST1\Data\NetworkAnalyst\DCStreets_ND_Junctions.shp, courtesy of US Census Bureau and Wilpen Gorr.

\EsriPress\GIST1\Data\NetworkAnalyst\FarmersMarkets.shp, courtesy of Wilpen Gorr.

\EsriPress\GIST1\Data\NetworkAnalyst\FQHC.shp, courtesy of Wilpen Gorr.

\EsriPress\GIST1\Data\NetworkAnalyst\PittsburghBlocks.shp, courtesy of US Census Bureau and Wilpen Gorr.

\EsriPress\GIST1\Data\NetworkAnalyst\PittsburghStreets.shp, courtesy of US Census Bureau and Wilpen Gorr.

\EsriPress\GIST1\Data\NetworkAnalyst\PittsburghStreets_ND_Junctions.shp, courtesy of US Census Bureau and Wilpen Gorr.

\EsriPress\GIST1\Data\NetworkAnalyst\SFStreets.shp, courtesy of US Census Bureau and Wilpen Gorr.

\EsriPress\GIST1\Data\NetworkAnalyst\XYNewLocs.shp, courtesy of Wilpen Gorr.

\EsriPress\GIST1\Data\NetworkAnalyst\SanFrancisco.gdb\Basemap, courtesy of NAVTEQ, TomTom, RATP.

\EsriPress\GIST1\Data\NetworkAnalyst\SanFrancisco.gdb\Transportation, courtesy of NAVTEQ, TomTom, RATP.

\EsriPress\GIST1\Data\NetworkAnalyst\SanFrancisco.gdb\DailyProfiles, courtesy of Wilpen Gorr.

\EsriPress\GIST1\Data\NetworkAnalyst\SanFrancisco.gdb\Kitchen, courtesy of Wilpen Gorr.

\EsriPress\GIST1\Data\NetworkAnalyst\SanFrancisco.gdb\Run21097, courtesy of Wilpen Gorr.

\EsriPress\GIST1\Data\NetworkAnalyst\SanFrancisco.gdb\SFGroceryStores, courtesy of Wilpen Gorr.

\EsriPress\GIST1\Data\NetworkAnalyst\SanFrancisco.gdb\SFGroceryWarehouses, courtesy of Wilpen Gorr.

EsriPress\GIST1\Data\NetworkAnalyst\SanFrancisco.gdb\Signposts_Streets, courtesy of NAVTEQ, TomTom, RATP.

EsriPress\GIST1\Data\NetworkAnalyst\SanFrancisco.gdb\Streets_DailyProfiles, courtesy of NAVTEQ, TomTom, RATP.

## MyExercises\FinishedExercises

\EsriPress\GIST1\MyExercises\FinishedExercises\Chapter5\Chapter5.gdb\tl_2010_17031_cousub10.zip, courtesy of US Census Bureau.

\EsriPress\GIST1\MyExercises\FinishedExercises\Chapter5\Topo.DWG, courtesy of Department of City Planning, City of Pittsburgh.

\EsriPress\GIST1\MyExercises\FinishedExercises\Chapter5\Chapter5.gdb\ACSLessThanHighSch, courtesy of US Census Bureau.

\EsriPress\GIST1\MyExercises\FinishedExercises\Chapter5\Chapter5.gdb\AfricanizedBees, courtesy of Data.gov.

\EsriPress\GIST1\MyExercises\FinishedExercises\Chapter5\Chapter5.gdb\Building, courtesy of Department of City Planning, City of Pittsburgh.

\EsriPress\GIST1\MyExercises\FinishedExercises\Chapter5\Chapter5.gdb\EcoRegions, courtesy of Data.gov.

\EsriPress\GIST1\MyExercises\FinishedExercises\Chapter5\Chapter5.gdb\Tornadoes, courtesy of Data.gov.

\EsriPress\GIST1\MyExercises\FinishedExercises\Chapter5\Chapter5.gdb\TractsPop2010, courtesy of US Census Bureau.

\EsriPress\GIST1\MyExercises\FinishedExercises\Chapter5\Chapter5.gdb\Chaptertl_2010_17031_areawater.zip, courtesy of US Census Bureau.

\EsriPress\GIST1\MyExercises\FinishedExercises\Chapter5\Chapter5.gdb\tl_2010_17031_cousub10.zip, courtesy of US Census Bureau.

\EsriPress\GIST1\MyExercises\FinishedExercises\Chapter5\Chapter5.gdb\tl_2010_17031_tract10.zip, courtesy of US Census Bureau.

\EsriPress\GIST1\MyExercises\FinishedExercises\Chapter5\DataGovShapefiles\afrbeep020.shp, courtesy of National Atlas of the United States, downloaded from NationalAtlas.gov (http://nationalatlas.gov/atlasftp.html#afrbeep).

\EsriPress\GIST1\MyExercises\FinishedExercises\Chapter5\DataGovShapefiles\ecoregp075.shp, courtesy of National Atlas of the United States, downloaded from NationalAtlas.gov (http://nationalatlas.gov/atlasftp.html#ecoregp).

\EsriPress\GIST1\MyExercises\FinishedExercises\Chapter5\DataGovShapefiles\tornadx020.shp, courtesy of National Atlas of the United States, downloaded from NationalAtlas.gov (http://www.nationalatlas.gov/atlasftp.html#tornadx).

\EsriPress\GIST1\MyExercises\FinishedExercises\Chapter5\ACS_10_5YR_S1501.zip, courtesy of US Census Bureau.

\EsriPress\GIST1\MyExercises\FinishedExercises\Chapter5\DEC_10_SF2_QTP1.zip, courtesy of US Census Bureau.

# Supplemental figure credits

## Chapter 3

Screen capture from ArcGIS Online, courtesy of i-cubed, information integration + imaging, LLC— distributed through i-cubed's DataDoors Archive Management www.datadoors.net.

Screen captures from ArcGIS Online, courtesy of National Agriculture Imagery Program, US Department of Agriculture, Farm Service Agency.

Screen captures from ArcGIS Online, courtesy of TomTom MultiNet.

Screen captures of www.census.gov, courtesy of US Census Bureau.

Screen captures of http://seamless.usgs.gov, courtesy of USGS.

## Chapter 5

Screen captures of http://nationalatlas.gov/mld/afrbeep.html, courtesy of Data.gov. Data.gov and the US federal government cannot vouch for the data or analyses derived from this data after the data has been retrieved from Data.gov.

Screen captures from ArcGIS Online courtesy of i-cubed, information integration + imaging, LLC—distributed through i-cubed's DataDoors Archive Management www.datadoors.net.

**Screen captures of ArcGlobe, data sources:**

Elevation (30m)—Source: USGS. The data is from the National Elevation Dataset (NED) produced by the United States Geological Survey (USGS).

Elevation (90m/1km)—Source: NASA, National Geospatial-Intelligence Agency (NGA), USGS. The data is from the National Elevation Dataset (NED) produced by the United States Geological Survey (USGS).

Copyright:© 2009 Esri, i-cubed, GeoEye elevation data includes 90 m Shuttle Radar Topography Mission (SRTM) elevation data from NASA and NGA where it is available and 1 km GTOPO30 data from the USGS elsewhere.

Imagery—Copyright:© 2009 Esri, i-cubed, GeoEye. This globe presents low-resolution imagery for the world and high-resolution imagery for the United States and other metropolitan areas around the world. The globe includes NASA Blue Marble: Next Generation 500 m resolution imagery at small scales (above 1:1,000,000), i-cubed 15 m eSAT imagery at medium-to-large scales (down to 1:70,000) for the world, and USGS 15 m Landsat imagery for Antarctica. It also includes 1 m i-cubed Nationwide Select imagery for the continental United States, and GeoEye IKONOS 1 m resolution imagery for Hawaii, parts of Alaska, and several hundred metropolitan areas around the world.

Boundaries and Places—Copyright:© 2009 Esri, AND, TANA. The map was developed by Esri using administrative and cities data from Esri and AND Mapping for the world and Tele Atlas administrative, cities, and landmark data for North America and Europe.

Transportation—Copyright:© 2009 Esri, AND, TANA. The map was developed by Esri using Esri highway data, National Geospatial-Intelligence Agency (NGA) airport data, AND road and railroad data for the world, and Tele Atlas.

# Data license agreement

Downloadable data that accompanies this book is covered by a license agreement that stipulates the terms of use.